Organic Analysis Using
Ion-selective Electrodes

THE ANALYSIS OF ORGANIC MATERIALS

An International Series of Monographs

edited by R. BELCHER and D. M. W. ANDERSON

Organic Analysis Using Ion-selective Electrodes

Volume 1 Methods

T. S. MA

*Department of Chemistry, City University
of New York, New York, USA*

and

S. S. M. HASSAN

*Department of Chemistry, Ain Shams University,
Cairo, Egypt*

1982

ACADEMIC PRESS

A Subsidiary of Harcourt Brace Jovanovich, Publishers

LONDON   NEW YORK

Paris   San Diego   San Francisco
São Paulo   Sydney   Tokyo   Toronto

ACADEMIC PRESS INC. (LONDON) LTD
24/28 Oval Road, London NW1 7DX

United States Edition published by
ACADEMIC PRESS INC.
111 Fifth Avenue, New York, New York, 10003

British Library Cataloguing in Publication Data

Ma, T. S.
Organic analysis using ion-selective electrodes.
—(Analysis of organic materials, ISSN 0309-2313)
Vol. 1
1. Chemistry, Analytic   2. Electrodes, Ion selective
3. Chemistry, Organic
I. Title   II. Hassan, S. S. M.   III. Series
547.3'08   QD271

ISBN 0-12-462901-6

Printed by
J. W. Arrowsmith Ltd., Bristol

PREFACE

Being commercially available and not expensive, ion-selective electrodes have become an item of general equipment for analytical work. During our recent visits to a number of chemical laboratories, however, we found that many such electrodes were not put to use either because the worker was disappointed with the erratic functioning of the electrodes or because he was unaware of their wide range of applicability. Therefore, we decided to prepare the present monograph in order to remedy these situations.

This book is organized in three parts. For convenience, it is published in two volumes. Volume 1 presents the background of ion-selective electrodes. The theoretical basis is discussed in a concise manner and the equipment and operations are briefly described. The characteristics of various types of ion-selective electrodes, as well as the different measuring techniques, are adequately treated. It may be mentioned that the reader who has some knowledge of ion-selective electrodes can glance through this portion of the book rapidly. On the other hand, this volume may serve as a textbook since it deals with the general aspects of ion-selective electrodes.

Volume 2, which comprises Parts Two and Three of the book, concentrates on the analysis of organic materials. Part Two involves an extensive survey of the literature on the use of ion-selective electrodes for organic analysis. The methods and procedures for the determination of the elements, various functional groups, biochemical substances, natural products and pharmaceuticals are critically reviewed. Potential applications of certain electrodes are also indicated. For the sake of clarity, many summaries are presented in tabular form.

Part Three presents 38 experiments as typical examples to cover the different areas of organic analysis. Detailed directions are given and each experiment is self-contained. These experiments have been verified by students in our laboratories.

By using this book, a practising chemist can learn the principles and methods of ion-selective electrodes, select a suitable procedure to analyse his samples, set up the appropriate equipment, carry out the experiment and check his results.

In recent years there have been spectacular developments in ion-selective electrodes and tremendous growth of their applications. The use of enzymes in conjunction with these sensors makes possible the selective determination of many complex biochemical systems. Current interest in microprocessors and automation has focused attention on the need for ion-selective electrodes owing to their simplicity, sensitivity and rapid response. It is hoped that the present monograph can serve as a guidebook and reference work in this important field.

We are indebted to Dr D. M. W. Anderson and Academic Press for their kindness in including this book in the Analysis of Organic Materials Series. We wish to acknowledge our appreciation to the numerous authors and publishers for their permission to reproduce illustrations, the original source being indicated under each figure. We thank C. Y. Wang of Yunnan University, Kunming, China for the detailed description of his home-made electrodes; his work demonstrates how one can perform organic analysis with ion-selective electrodes when there is no money to purchase commercial models. We are grateful to Dr Amina Ads (Mrs S. S. M. Hassan) for her valuable assistance in the preparation of the manuscript, and Mrs Mei-Mei M. Hewitt for secretarial service.

New York   *T. S. Ma*
October 1981   *S. S. M. Hassan*

CONTENTS OF VOLUME 1

CONTENTS OF VOLUME 2

EXPERIMENTAL PROCEDURES

METHODOLOGY

1. THEORETICAL AND PRACTICAL ASPECTS OF ION-SELECTIVE ELECTRODES

I. MECHANISM OF OPERATION

A. Origin of membrane potential

Generally speaking, ion-selective electrodes [1, 2] consist of electrochemical membranes which are composed of either a solid or a liquid phase permeable only to one ion species. The properties of the membrane phase are dependent on its composition and the velocity of permeation to the various ions. These electrodes utilize an internal solution of constant composition in contact with the membrane and an internal reference electrode. When the ion-selective electrode is placed in a solution containing the particular ion (to which the electrode is reversible), a small number of ions (to which the membrane is selective) pass from the solution of higher concentration through the membrane to that of lower concentration, thus producing an electric potential difference [3] known as "liquid junction potential" or "diffusion potential" [3a–c]. Consequently the transmembrane difference of electrode potential (Donnan potential) [4–6] is generated which eventually stops further diffusion of the particular ion species. An electric double layer is simultaneously formed on both sides of the membrane (see Fig. 1.1).

The equilibrium between the solution and the membrane phase is attained when the difference of the electrochemical potentials between the solvated ion (n_i) and the ion bonded to the membrane (n_s) equals zero. The ion diffusion across the membrane is not considered; it normally has no effect on the membrane potential at zero current [7].

Therefore the electrochemical potential in the solution is:

$$(\eta_i) = \mu_i + Z F \Psi_i \tag{1}$$

Numbers in square brackets refer to numbered references at the end of each chapter.

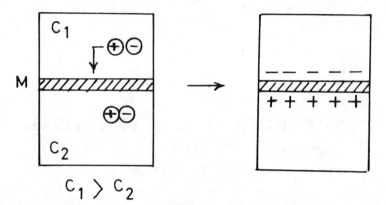

FIG. 1.1. Diagram of an electrochemical membrane (M) selective to the cationic species.

and the electrochemical potential across the membrane is:

$$(\eta_s) = \mu_s + ZF\Psi_s \tag{2}$$

where: η is the electrochemical potential;
μ is the chemical potential;
Z is the valency of the test ion;
F is the Faraday constant;
Ψ is the Galvani potential.

At equilibrium $\eta_i = \eta_s$, and

$$\mu_i + ZF\Psi_i = \mu_s + ZF\Psi_s \tag{3}$$

$$ZF(\Psi_s - \Psi_i)\mu_i - \mu_s = \mu_i^0 - \mu_s^0 + RT \ln (a)_i/(a)_s \tag{4}$$

μ^0 being the standard chemical potential. Equation 4 then becomes:

$$E = E_0 + RT/ZF \ln (a)_i/(a)_s \tag{5}$$

where: E is the electrode potential;
E_0 is the standard electrode potential;
a_i and a_s are the respective activities of the ion in solution and in the membrane phase.

Since the electrode contains an internal solution of constant composition, the electrical potential measured using such a half cell depends solely on the activity of the ion in the external test solution. Therefore:

$$E = E_0 + RT/ZF \ln a_i \tag{6}$$

All the various junction potentials participating in the net e.m.f. are taken as constant except that which depends on the nature of the test solution.

B. Nature of the membranes

While the potential of the membranes can be explained by taking into consideration both the diffusion and Donnan potentials, other factors including boundary conditions appropriate to ion exchangers may also be involved [8]. Membranes can thus be classified on this basis into the following categories.

1. Solid State Membranes

These membranes comprise a nonporous layer of a sparingly soluble salt (compact poly crystalline or single crystal) in contact with a solution containing a single ion species. The membrane may be formed from more than one component. Thus, most metal sulfides, other than silver sulfide, can only be used as active material if mixed with the latter. Silver sulfide is also used in silver halide membranes. For the silver halide membrane, the potential depends on the anion activity in the solution, while the silver ion serves as the charge carrier in the membrane. In some membranes of this type, a chemical reaction may take place under special activity ratios and selectivity parameters (P). Thus when the internal solution of the silver halide electrode contains chloride ions only, while the test solution contains both chloride and iodide ions, if the iodide activity $a_I < (P_{AgI}/P_{AgCl})a_{Cl}$, the presence of iodide does not affect the membrane potential; but if $a_I > (P_{AgI}/P_{AgCl})a_{Cl}$ then the silver chloride in the surface region of the membrane is converted to silver iodide [9].

2. Solid Ion-exchanger Membranes

In the compact ion-exchanger membranes, an ion-exchange reaction takes place between the ions in solution and those in the membrane [10–12] according to the equation:

$$J_m^+ + K_s^+ \rightleftharpoons J_s^+ + K_m^+$$

The relative mobilities of the ionic species K^+ or J^+ in the membrane and the exchange equilibrium constant K_{kj} affect the potential of the membrane.

When such membranes are brought into contact with an electrolyte containing the two ion species J^+ and K^+, ion exchange in the membrane surface takes place, and a constant concentration of these species is soon established in the solution rather than in the membrane since diffusion inside the membrane is much slower than in the solution. The selectivity of the membrane [13], therefore, depends on the selectivity constant for K^+ with respect to that for J^+ ($K_{Jk}^{pot} = 1/K_{kj}^{pot}$). If $a_k \gg K_{kj}^{pot} a_j$, the membrane responds to K^+ in a Nernstian manner; but if $a_k \ll K_{kj}^{pot} a_j$, the membrane responds to J^+.

3. Liquid Ion-exchanger Membranes

In these membranes, anion or cation exchangers are dissolved in a water-immiscible solvent. When these membranes are brought into contact with solutions containing J^+ and K^+ ions, association with the exchanger site in the membrane takes place [14–17].

For a liquid ion-exchanger, in which association between counter ion and site cannot be neglected, the origin of selectivity is generally a function of both the solvent and the exchanger. For a strongly dissociated liquid ion-exchanger, the selectivity is completely independent of the chemical properties of the ion-exchanger but not of the sign of its charge. For example, nitrobenzene when dissolved in some carboxylic, sulfonic or phosphoric acids can be used as a cation-selective membrane, but when dissolved in high amines (e.g. dodecylamine) it serves an an anion-selective membrane with high selectivity towards halogens [14].

4. Neutral Membranes

Neutral molecules of some antibiotics and polycyclic polyethers have been found to produce cation-selective membrane electrodes. High selectivity of such systems has been reported for nonactine in carbon tetrachloride, valinomycin in hexane and valinomycin in octanol [18–21]. All these substances contain no charged functions but an arrangement of ring oxygens energetically suitable to replace the hydrated shell around cations. Thus, these soluble molecules are able to solubilize cations in organic solvents forming mobile charged complexes with the cations therein. In this way a mechanism for cationic permeation across such media is provided.

Recently, ion-selective disk electrodes with neutral cation-carriers were described by Schindler et al. [21a]. The electrodes incorporate synthetic electrically neutral carrier molecules for Na^+ and Ca^{2+}, and valinomycin as carrier for K^+.

C. Selectivity of membranes

When the test solution contains ions other than those to which the electrode is essentially reversible, a change in the electrical potentials depends upon the selectivity of the membrane. For an electrode selectively responding to an ion (i) of activity (a_i) and charge (Z) in the presence of an interfering ion (j) of activity (a_j) and charge (Y), the e.m.f. is given by equation 7:

$$E = E_0 + RT/ZF \ln [a_i + K_{ij}(a_j)^{Z/Y}] \tag{7}$$

where K_{ij} is the selectivity constant. For response to the ion (i) only, K_{ij} must be small. This constant is related to an ion-exchange process at the electrode surface and its value, calculated from the empirical equation, can be used as an index of selectivity for all ion-selective electrodes.

If the solution not only contains the ion to which the electrode is reversible, but also another ion which forms a precipitate or a complex with one of the constituents of the membrane matrix, precipitation or a complex exchange reaction takes place [22, 23].

$$Agi_{(m)} + j_{(s)} \rightleftharpoons Agj_{(m)} + i_{(s)}$$

$$(i = Cl^-, j = I^-, i = CNS^-, j = Br^-, i = I^-, j = CN^-) [22-25]$$

$$K_{ij} = (a_i)_s(a_j)_m/(a_j)_s(a_i)_m \tag{8}$$

Consequently, the potential of the membrane electrode can be given on the basis of these exchange equilibria and by neglecting the intra-membrane diffusion phenomena as follows:

$$E = E_0 + RT/Z_iF \ln (a_i)_s \sum_n ij(a_j)_s/(a_i)_s \tag{9}$$

where (a_i) and (a_j) are the activities of (i) and (j) ions, respectively, n is the number of ions taking part in the exchange reaction, K_{ij} is the selectivity constant of the electrode.

The selectivity constant can be deduced from the exchange equilibrium and has the following general forms:

$$K_{ij} = S_{ji}^{1/a}/S_p^{1/n} = e(a_i)_s^{b/a}/e(a_j)_s^{m/n} \tag{10}$$

where $e(a_i)$ and $e(a_j)$ are equilibrium activities of the ions taking part in the exchange reaction, a, b, n, m are the stoichiometric constants of the precipitates built in the membrane or formed during the exchange reaction, and S_{ji} and S_p are the appropriate solubility products. When univalent ions form the precipitates, equation 10 becomes:

$$K_{ij} = S_{ji}/S_p \tag{11}$$

However, the selectivity ratio can be experimentally evaluated using various procedures [26, 27]. One of these involves potential measurement of the solution containing the pure primary ion (i), and for that containing the interfering ion (j). The potential of the electrode in such a solution can be expressed using equation 7 and substituting for the value of a_i by 0. If the ions to be measured are univalent anions, then equation 7 becomes:

$$E_2 = E_1 - 2 \cdot 303RT/F \log K - 2 \cdot 303RT/F \log a_j \tag{12}$$

If $a_i = a_j$, then

$$\log K = (E_1 - E_2)/2 \cdot 303RT/F \tag{13}$$

Measurements of the potential of the electrode in a solution containing both ions can also be utilized. By combining equations 7 and 12, one obtains:

$$E_1 - E = 2 \cdot 303RT/F \log (a_i + Ka_j)/a_i \tag{14}$$

For expressing K, the following equation is obtained.

$$K = \frac{\left(\exp\left\{\dfrac{E_1 - E}{RT/F}\right\}\right)a_i - a_i}{a_j} \tag{15}$$

Equation 15 can only be used for the calculation of the selectivity ratio from two potential measurements, one in a pure solution of the ion (i) and the other in a mixture of the ions (i) and (j). However, a graphical procedure involving a series of measurements can be employed by measuring E_1 in a known volume of the solution containing the ion (i) at an activity (a_i), then known volumes of a stock solution containing the ion (j) are added successively followed by potential measurements after each addition (E'). As a result, a series of values for E' is obtained, each value corresponding to a set of values for the activities of the ions (i) and (j) in the same solution. If the potential measured in any one of the series of solutions is E', then from equations 7 and 12:

$$\left[\exp\left\{\frac{E_1 - E'}{RT/F}\right\}\right]a_i - a_i' = Ka_j' \tag{16}$$

By plotting the left-hand side of equation 16 against a_j', a straight line passing through the origin, with a slope corresponding to the selectivity ratio, is obtained. Also, the following equation can be used instead.

$$a_i' - \left[\exp\left\{\frac{E - E'}{RT/F}\right\}\right]a_i = K\left[\left(\exp\left\{\frac{E - E'}{RT/F}\right\}\right)\right]a_j - a_j' \tag{17}$$

Then by plotting the left-hand side of equation 17 against the function within the brackets on the right-hand side, a straight line passing through the origin with a slope corresponding to the selectivity ratio is obtained [27]. Mechanistic information concerning the selectivity of several ion-selective electrodes has been obtained by investigations based on exchange–current measurements carried out under varying solution conditions [28].

Baumann [28a] recently described a coated-platinum sulfate selective electrode which was made by coating a platinum electrode with a mixture of aliquat-336 (SO_4^{2-} form) and 4'-butyl-2,2,2-trifluoroacetophenone in a poly(vinyl chloride) matrix. Addition of 2-aminopyrimidine sulfate or barium sulfate to the mixture was found to improve the selectivity for SO_4^{2-} relative to NO_3^-.

It may be noted that solid-state membranes are more selective than liquid membranes, that most anions do not interfere with the cation-selective membranes and vice versa, and that information on the selectivity of membranes in the presence of more than one interfering ion is scanty.

The selectivity constants of the various ions with some commercially available liquid and solid membrane electrodes are given in Table 1.1.

Table 1.1. Selectivity constants for some commercially available electrodes

Electrode	Type and model	State	Selectivity constant
Boron tetra-fluoride	Orion 92-05	Liquid	F^- 10^{-3}, Cl^- 10^{-3}, Br^- 0.04, I^- 20, NO_3^- $0\cdot1$, SO_4^{--} 10^{-3}, HCO_3^- 4×10^{-3}, CH_3COO^- 4×10^{-3}, OH^- 10^{-3}
Boron tetra-fluoride	Beckman 39620	Liquid	F^- 2×10^{-4}, Cl^- 5×10^{-4}, Br^- $0\cdot02$, I^- $0\cdot13$, NO_3^- $0\cdot02$, SO_4^{--} $<10^{-6}$, S^{--} 10^{-4}, CO_3^- 6×10^{-6}, CH_3COO^- $1\cdot5\times10^{-4}$, ClO_3^- $0\cdot03$, CN^- 6×10^{-4}, PO_4^{3-} 2×10^{-4}, $[Fe(CN)_6]^{3-}$ $<10^{-6}$
Bromide	Beckman 39602	Solid	Cl^- 3×10^2, I^- $1\cdot8\times10^{-4}$, OH^- $4\cdot4\times10^4$, CN^- $4\cdot1\times10^{-4}$, SCN^- $1\cdot8$
Bromide	Coleman 3-801	Solid	Cl^- 400, I^- 2×10^{-4}, OH^- 3×10^{-4}
Bromide	Orion 94-53	Solid	Cl^- 400, I^- 2×10^{-4}, OH^- 3×10^{-4}, CN^- 8×10^{-5}
Bromide	Radelkis OP-I-711	Solid	Cl^- 200, I^- $7\cdot7\times10^{-3}$
Bromide	Philips IS-550	Solid	Cl^- 6×10^{-3}, I^- 20, OH^-, 10^{-3} CN^- 25, CO_3^- $2\cdot3\times10^{-3}$, $S_2O_3^-$ $1\cdot5$, CrO_4^{-2} $1\cdot6\times10^{-3}$
Calcium	Beckman 39608	Liquid	Mg^{++} $0\cdot11$, Ba^{++} $0\cdot08$, Sr^{++} $0\cdot09$, Na^+ $0\cdot015$, H^+ 70, K^+ 0.034, Cd^{++} 3, Mn^{++} 4, Cu^{++} 3, Fe^{++} $0\cdot2$
Calcium	Orion 92-20	Liquid	Zn^{++} $3\cdot2$, Fe^{++} $0\cdot8$, Pb^{++} 63, Mg^{++} $0\cdot01$, Ba^{++} $0\cdot01$, Sr^{++} $0\cdot017$, Ni^{++} $0\cdot08$, Cu^{++} $0\cdot27$, Na^+ $1\cdot6\times10^{-3}$, K^+ 10^{-4}, NH_4^+ 10^{-4}, H^+ 10^5
Calcium	Corning 476041	Liquid	Mg^{++} $0\cdot01$, Ba^{++} $0\cdot01$, Sr^{++} $0\cdot01$, Ni^{++} $0\cdot01$, Na^+ 10^{-3}, K^+ 10^{-3}
Calcium/magnesium (Water hardness)	Orion 92-32	Liquid	Zn^{++} $3\cdot5$, Fe^{++} $3\cdot5$, Ba^{++} $0\cdot94$, Sr^{++} $0\cdot54$, Ni^{++} $1\cdot35$, Cu^{++} $3\cdot1$, Na^+ $0\cdot01$, K^+ $<0\cdot015$
Calcium/magnesium (Water hardness)	Beckman 39614	Liquid	Zn^{++} $>0\cdot1$, Mg^{++} $0\cdot95$, Ba^{++} $0\cdot8$, Na^+ $0\cdot013$, K^+ $0\cdot013$

Table 1.1. (cont.)

Electrode	Type and model	State	Selectivity constant
Chloride	Beckman 39604	Solid	$Br^- 3 \times 10^{-3}, I^- 5 \times 10^{-7}, OH^- 80, CN^- 2 \times 10^{-7}$
Chloride	Coleman 3-802	Solid	$Br^- 4 \cdot 9 \times 10^{-3}, I^- 10^{-6}, OH^- 100, S_2O_3^- 0 \cdot 01$
Chloride	Orion 94-17	Solid	$Br^- 3 \times 10^{-3}, I^- 5 \times 10^{-7}, OH^- 80, CN^- 2 \times 10^{-7}$
Chloride	Philips IS-550	Solid	$Br^- 1 \cdot 2, I^- 86 \cdot 5, OH^- 0 \cdot 024, CN^- 400, CO_3^{--} 3 \times 10^{-3}, S_2O_3^{--} 0 \cdot 01, CrO_4^{--} 1 \cdot 8 \times 10^{-3}$
Chloride	Radelkis OP-Cl	Solid	$Br^- 4 \cdot 95 \times 10^{-3}, I^- 2 \cdot 8 \times 10^{-3}, SO_4^{--} 4 \cdot 95 \times 10^5$
Chloride	Corning 476131	Liquid	$Br^- 2 \cdot 5, I^- 15, NO_3^- 2 \cdot 5, CH_3COO^- 0 \cdot 21, ClO_4^- 5, OH^- 0 \cdot 4$
Chloride	Orion 92-17	Liquid	$F^- 0 \cdot 1, Br^- 1 \cdot 6, I^- 17, NO_3^- 4.2, SO_4^{--} 0 \cdot 14, HCO_3^- 0 \cdot 19, CH_3COO^- 0 \cdot 32, ClO_4^- 32, OH^- 1 \cdot 0$
Copper	Orion 92-29	Solid	$Zn^{++} 10^{-3}, Fe^{++} 1 \cdot 0, Mg^{++} 10^{-4}, Ba^{++} 10^{-4}, Sr^{++} 10^{-4}, Ni^{++} 5 \times 10^{-3}, Ca^{++} 5 \times 10^{-4}, Na^+ < 10^{-5}, K^+ < 10^{-3}, H^+ 10$
Cyanide	Philips IS550-CN	Solid	$Cl^- 2 \times 10^{-4}, Br^- 2 \times 10^{-5}, I^- 3, CO_3^{--} 3 \cdot 6 \times 10^{-4}, S_2O_3^{--} 2 \cdot 2 \times 10^{-3}, CrO_4^{--} 1 \cdot 4 \times 10^{-2}$
Cyanide	Orion 94-06	Solid	$Cl^- 10^6, Br^- 5 \times 10^{-3}, I^- 0 \cdot 1$
Cyanide	Radelkis OP-CN	Solid	$Cl^- 1 \cdot 7 \times 10^5, Br^- 210, SO_4^{--} 3 \cdot 2 \times 10^7, PO_4^{---} 4 \cdot 8 \times 10^5, ClO_4^- 1 \cdot 6 \times 10^6$
Iodide	Beckman 39606	Solid	$Cl^- 1 \cdot 6 \times 10^6, Br^- 5 \cdot 6 \times 10^{-3}, CN^- 0 \cdot 4, S_2O_3^{--} 10^5, SCN^- 10^4$
Iodide	Orion 94-53	Solid	$Cl^- 10^6, Br^- 5 \times 10^3, CN^- 0 \cdot 4, S_2O_3^{--} 10^5$

Table 1.1. (cont.)

Electrode	Type and model	State	Selectivity constant
Iodide	Radelkis OP-I	Solid	Cl^- $1{\cdot}7 \times 10^5$, Br^- 210, SO_4^{--} $3{\cdot}2 \times 10^7$, PO_4^{---} $4{\cdot}8 \times 10^5$, ClO_4^- $1{\cdot}6 \times 10^6$
Iodide	Philips IS-550-I	Solid	Cl^- $6{\cdot}6 \times 10^{-6}$, Br^- $6{\cdot}5 \times 10^{-5}$, CN^- $0{\cdot}34$, CO_3^{--} $1{\cdot}2 \times 10^{-4}$, $S_2O_3^{--}$ $7{\cdot}1 \times 10^{-4}$, CrO_4^{--} $3{\cdot}7 \times 10^{-3}$
Lead	Orion 92-82	Solid	Zn^{++} 3×10^{-3}, Fe^{++} $0{\cdot}08$, Mg^{++} 8×10^{-3}, Ni^{++} 7×10^{-3}, Cu^{++} $2{\cdot}6$
Nitrate	Corning 476134	Liquid	Cl^- 4×10^{-3}, Br^- $0{\cdot}011$, I^- 25, SO_4^{--} 10^{-3}, HCO_3^- 10^{-3}, CH_3COO^- 10^{-3}, ClO_4^- 10^3
Nitrate	Orion 92-07	Liquid	F^- 6×10^{-5}, Cl^- 6×10^{-3}, Br^- $0{\cdot}9$, I^- 20, NO_2^- $0{\cdot}06$, SO_3^{--} 6×10^{-3}, SO_4^{--} 6×10^{-3}, S_2O_3 6×10^{-3}, S^{--} $0{\cdot}57$, HS^- $0{\cdot}04$, CO_3^{--} 6×10^{-3}, HCO_3^- $0{\cdot}02$, CH_3COO^- 6×10^{-3}, ClO_3^- 2, ClO_4^- 10^3, CN^- $0{\cdot}02$, PO_4^{---} 3×10^{-4}, $H_2PO_4^-$ 3×10^{-4}, HPO_4^{--} 8×10^{-5}
Nitrate	Beckman 39618	Liquid	F^- $6{\cdot}6 \times 10^{-3}$, Cl^- $0{\cdot}02$, Br^- $0{\cdot}28$, I^- $5{\cdot}6$, NO_2^- $0{\cdot}066$, SO_4^{--} 10^{-5}, S^{--} $3{\cdot}5 \times 10^{-3}$, CO_3^{--} $1{\cdot}9 \times 10^{-4}$, CH_3COO^- 5×10^{-3}, ClO_3^- $1{\cdot}1$, ClO_4^- $95{\cdot}5$, CN^- $0{\cdot}02$, PO_4^{---} $7{\cdot}4 \times 10^{-3}$
Perchlorate	Orion 92-81	Liquid	F^- $2{\cdot}5 \times 10^{-4}$, Cl^- $2{\cdot}2 \times 10^{-4}$, Br^- $5{\cdot}6 \times 10^{-4}$, I^- $0{\cdot}012$, NO_3^- $1{\cdot}5 \times 10^{-3}$, SO_4^{--} $1{\cdot}6 \times 10^{-4}$, HCO_3^- $3{\cdot}5 \times 10^{-4}$, CH_3COO^- $5{\cdot}1 \times 10^{-4}$, OH^- $1{\cdot}0$
Perchlorate	Beckman 39616	Liquid	F^- 10^{-4}, Cl^- 10^{-4}, Br^- 3×10^{-3}, I^- $0{\cdot}04$, NO_3^- $6{\cdot}6 \times 10^{-3}$, SO_4^{--} $<10^{-6}$, S^{--} 5×10^{-5}, CO_3^{--} 2×10^{-6}, CH_3COO^- 5×10^{-5}, ClO_3^- $0{\cdot}01$, CN^- 2×10^{-4}, PO_4^{---} 10^{-4}
Sulfide	Orion 94-16	Solid	CN^-, CO_3^{--} HCO_3^-, SO_4^{--}, SO_3^{--}, $S_2O_3^{--}$, F^-, Cl^-, Br^-, I^- $< 10^{-3}$

D. Activity and concentration

Whereas the electrode potential follows a Nernstian relation expressing the activity (equation 6), it is necessary to discuss the relationship between activity and concentration since many analyses are performed for concentration measurements.

The relationship between the activity and concentration of an ion can be expressed by the equation

$$[A^\pm] = \gamma_{A^\pm} \times m_{A^\pm} \tag{18}$$

where $\gamma_{A^\pm}$ is the practical activity coefficient of $A^\pm$ and $m_{A^\pm}$ is its molality. In dilute solutions at room temperature, molarity and molality are approximately equal, so that

$$[A^\pm] = \gamma_{A^\pm} \times (A^\pm) \tag{19}$$

Values for a single ion activity coefficient $\gamma_{A^\pm}$ cannot be measured, but may be calculated approximately from equation 20 based on Debye–Hückel theory [29].

$$\log \gamma_A = -AZ^2\sqrt{\mu} \tag{20}$$

where Z is the charge of the ion, μ the ionic strength of the solution and A is a constant which depends on temperature and solvent.

Since the Debye–Hückel equation is correct in the limit of zero ionic strength, the activity coefficient of an ion will depend only on its charge and the ionic strength of the environment. Consequently, as the ionic strength (μ) increases, the activity coefficient (γ) decreases.

The total ionic strength of a solution is given by equation 21.

$$\mu = \tfrac{1}{2}\sum Z_i^2 C_i \tag{21}$$

i.e. the concentration of each ion species in the solution (C_i) is multiplied by the corresponding square of the charge on the ion (Z_i), and the sum of these products is divided by 2.

For example, a solution of the following composition:

0·15 mol/liter of NaCN
0·10 mol/liter of K_2SO_4
0·15 mol/liter of NaOH
0·10 mol/liter of $NaNO_3$

has a total ionic strength calculated as shown in Table 1.2. Therefore, the total ionic strength (μ) = 1·4/2 = 0·70 M.

If the approximate sample composition is known, the total ionic strength can be estimated as shown above. By using Fig. 1.2, the activity coefficient γ can be derived, and consequently the concentration. A nomogram for

Table 1.2

Ion	Z_i^2	C_i	$Z_i^2 C_i$
Na^+	1	$\times 0.15 + 0.15 + 0.10$	$= 0.40$
K^+	1	$\times 0.10 + 0.10$	$= 0.20$
SO_4^{2-}	4	$\times 0.10$	$= 0.40$
OH^-	1	$\times 0.15$	$= 0.15$
CN^-	1	$\times 0.15$	$= 0.15$
NO_3^-	1	$\times 0.10$	$= 0.10$
		Total	1.40

converting concentration of ions in mg/l to their contributions to ionic strength is available [30].

It is interesting to note that, while the electrical potential is related to ion activity, conductance is related to ion concentration. Thus Powley *et al.* [30a] made dipolar pulse conductance measurements with a calcium ion-selective electrode and found that the conductance is a function of calcium concentration and not of activity.

II. GENERAL PROPERTIES OF ION-SELECTIVE ELECTRODES

A. Response-speed and stability

The response time of an ion-selective electrode is the time needed to attain an equilibrium value within $\pm 1\,mV$ after a ten-fold increase or decrease

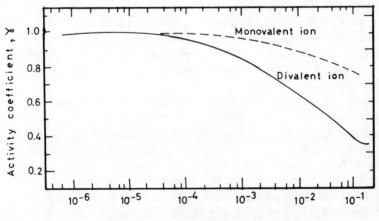

FIG. 1.2. Ion activity coefficient as a function of total ionic strength.

in the concentration of the test solution. This time may be obtained by immersing the electrode in the test solution and measuring the time needed to attain a constant potential either by using a chart recorder, or more accurately an oscilloscope to measure the time in the millisecond range. Sensors with fast response are used in automatic analyzers, continuous measurement cells, kinetic studies and potentiometric titrations.

1. *Response of Solid State and Liquid Membrane Electrodes*

The speed and stability of the electrodes depend on several factors such as: (i) the type of electrode membrane; (ii) the condition of the membrane surface; (iii) the stirring rate; (iv) the concentration and viscosity of the test solution; (v) the presence of interfering ions; and (vi) magnitude and direction of concentration change.

A number of ion-selective electrodes have been subjected to extensive investigation [7, 31–35a]. Some findings are briefly discussed below.

Study of the homogeneous solid state fluoride ion-selective electrode shows that the electrode has good time response (92% response in 1 min or less for fluoride concentration above 10^{-6} M) [36]. A drift of ± 2 mV–7 mV is observed depending on stirring and activity conditions. At high ionic strength, the response is more stable [37–39]. This drift is attributed to temperature changes and to the variability of the liquid junction potential with the reference electrode. Stirring the solution has a substantial effect on the observed potential [40]. In 10^{-3} M sodium fluoride solution, the potential changes from $-61 \cdot 5$ mV in a quiescent solution to $-55 \cdot 5$ mV in a rapidly stirred solution. This shift is less at high concentration and negligible in the presence of 1 M sodium nitrate supporting electrolyte. It is recommended, therefore, to take the readings with slow stirring, when the potential is quite stable and the change in it does not exceed $0 \cdot 1$ mV even after the lapse of one hour. It was also observed that the change in the potential reading of the fluoride ion-selective electrode after 2 weeks was 3 mV more negative, and after a month 7 mV more negative for the same solution. The potential of the fluoride microelectrode varies by about 20 mV from day to day [41].

The time response of heterogeneous solid state electrodes with silicone rubber matrix (Pungor type) such as iodide, bromide, and chloride electrodes are 8, 14 and 20 s, respectively. This may indicate that the e.m.f. developed by the electrode arises from an ion exchange mechanism at the surface. The response of these types of electrodes [31] is dependent on concentration within 10^{-2} and 10^{-1} M.

Heterogeneous solid state electrodes with poly(vinyl chloride) matrix take less than 2 min to reach equilibrium, but the presence of interfering ions may significantly affect the response time [42]. For example when the nitrate poly(vinyl chloride) electrode is used for measuring the nitrate ion in the presence of perchlorate, it takes 40 min before returning to normal behavior with pure nitrate solution. The same behavior is also observed when the calcium poly(vinyl chloride) electrode is exposed to zinc solution.

Heterogeneous solid state membrane-coated wire electrodes with poly (vinyl chloride) [43] or poly(methyl methacrylate) [44] matrix show a response time of 30–60 s with reproducibility within ±1 mV, but the potential readings vary from day to day between 5–15 mV. The effect of surface tarnishing on the response of sulfide ion-selective electrodes has been studied [45].

Liquid ion-selective electrodes show a potential response varying from seconds to minutes. The potassium electrode response [46, 47] is 1 min at the 10^{-1} M potassium level, and 10–15 min for 10^{-4}–10^{-6} M. The average response time of the calcium electrode is 10 s [48]. It responds more slowly to change in calcium activity when magnesium ion is present. The electrode for nitrate determination [48a] responds in a few seconds to NO_3^- solution of 1–100 mM, but requires 3 min at 10 μM. In general, electrodes of different types are liable to different extents of drift. Some electrodes (e.g. most of the liquid ion-selective electrodes) drift 2 mV in one day and necessitate frequent recalibration, whereas other electrodes have very low drift such as the potassium valinomycin membrane electrode which does not drift before ten days [49].

With all types of ion-selective electrodes [49–51] the following points may be observed:

(a) Steady potential can be read out depending on the rate of stirring, especially at low activities. Differences in potential response between stirred and unstirred solution, and for different rates of stirring may occur. The effect of stirring is very pronounced with all types of liquid ion-selective electrodes and some solid state electrodes such as the cyanide electrode [25].

(b) Several seconds are needed for various electrodes in 1 M ion solution, whereas several minutes are needed at the limit of electrode detection.

(c) The electrode response is more rapid when going from dilute to concentrated solution, with the exception of the sodium electrode.

(d) The speed of response and stability of the potential are much better with solid state membrane electrodes than with the liquid membrane type electrodes.

2. Response of Enzyme Electrodes

The enzyme electrodes are either solid state membrane electrodes or gas sensing membrane probes with immobilized enzymes. These electrodes sense either ionic or gaseous species as a result of the reaction of the enzyme with a suitable substrate. The response of the enzyme electrodes depends on: (i) the nature and response behavior of the electrode used, (ii) the nature and sensitivity of the enzyme reaction, (iii) the concentration of both the enzyme and substrate, and (iv) the operation temperature.

In general, the enzyme electrodes exhibit a response time ranging from a few seconds to a few minutes. The response time of the penicillin electrode [52] is 15–30 s, whereas the urea electrode [53] attains 98% of the steady state response in 60–180 s, depending on the enzyme and substrate concentrations.

3. *Response of Cation-sensitive Glass Electrodes and Transient Phenomena*
Many of the cation glass electrodes respond within a few minutes, and their potential undergoes a momentary (~100 ms) excursion far beyond the expected when these electrodes are subjected to a sudden change in the concentration of monovalent cations in a system of constant concentration of monovalent cations [54, 55] (Fig. 1.3). This "transient" phenomenon is not observed on the normal laboratory time scale, but may be observed when the electrode potential is monitored as a function of time in a flowing system. A similar effect is also observed with liquid membrane and solid state electrodes [56–60].

4. *Time Response Paper*
This paper (Fig. 1.4) was developed by Orion [61] to allow the electrode time response to be expressed as a number and to read the potential of slow electrode response after a short time from a predicted time equals infinity mV reading. The horizontal axis on this paper is the reciprocal of time while the vertical is the antilog electrode potential/slope. The straight line drawn through the plotted points intercepts the ordinate at the left of the paper to give the time equals infinity potential.

The time response paper is used in such a manner that for mono- and divalent electrodes each major division is equal to 5 and 2·5 mV respectively, and for cation and anion electrodes increasingly positive and negative potentials, respectively, may be plotted up the vertical axis. The time axis can be scaled as required; for example, each division may represent 0·1 min or 10 s.

B. Life-span

The life-span of ion-selective electrodes depends on: (i) the type of the electrode and the nature of its membrane, (ii) operation temperature, (iii) concentration of the measuring ion, (iv) presence of interfering or corrosive substances, and (v) operation condition or technique (e.g. in a flowing or closed system).

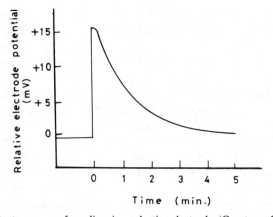

FIG. 1.3. Transient response of a sodium ion-selective electrode. (Courtesy of Orion Research Inc.)

The commercially available solid state electrodes have a life-span of one to three years when used in dilute solutions under normal conditions and one to three months at high temperatures or in a flowing system. It was reported that the chloride electrode (Orion 94-17) can be used for the analysis of at least 20,000 samples [62] and the bromide electrode (Orion 94-35) can be used for potentiometric titration of the chloride ion with mercuric ions without a noticeable deterioration [63] in response. Beckman electrodes have been designed with an easily replaceable sensor element tip which can be used for at least 100 measuring hours before replacing

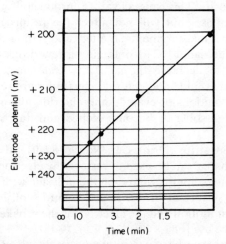

FIG. 1.4. Electrode time response paper (according to *Orion Newsletter* **3**, 11 (1971)). (Courtesy of Orion Research Inc.)

with a new solid element unit and internal reference solution. However, the life of the cyanide electrode (AgI/Ag_2S) is shorter because cyanide ion destroys the electrode membrane, and this is a serious drawback of the electrode. Its estimated lifetime [64], at 10^{-3} M is 200 h but only about 20 h at 10^{-2} M. Since 1 min is sufficient for measuring a cyanide sample, 20 h means the possibility of analyzing hundreds of cyanide sample solutions. This drawback can be obviated if the indicator measurement technique using the silver sulfide electrode is adopted.

Liquid ion-selective electrodes can be used for one to three months in continuous service (except the potassium electrode which is only used for 2–3 weeks) without renewal of the liquid ion-exchanger. This short lifetime may be related to the gradual loss of the ion exchanger through the porous membrane. The membrane, internal filling solution and the ion exchanger are replaced when the electrode response becomes noisy or drifts. In general, the liquid membrane electrodes and the gas sensing membrane probes are good for use through a period of one month or longer if no mechanical problems such as leaking develops [65]. The enzyme electrodes can be used continuously for 10–12 h at or above the room temperature without loss of activity [66].

C. Conditioning and rejuvenation

Conditioning of homogeneous solid state ion-selective electrodes can be conducted by (i) polishing with special polishing paper, (ii) wiping with a toothbrush and toothpaste, (iii) treatment with chemical reagents, or (iv) soaking in a standardized ion solution.

The membrane of a lead electrode tends to passivate with time due to accumulated deposits. This causes loss of reproducibility to about 1–3 mV and sluggish response, but will not affect the results of potentiometric titration. For direct measurements, the test and standard solutions should be diluted with methanol [67] (containing a few drops of formalin/liter) in the ratio 1:1 to decrease the solubility and retard oxidation of the membrane. When cyanide and chloride electrodes are repeatedly exposed to concentrated solutions of cyanide and chloride ions respectively, their membranes may dissolve or become etched and dull, and the response deteriorates.

Special polishing strips (Orion Cat. No. 94-82-01), tissue paper or 3/0 emery polishing paper can be used to restore the electrode condition. One inch of the polishing paper is wetted on the dull side with one or two drops of water and the surface of the sensing element is polished for about 30 s by an even circular motion and the electrode is then soaked in a standardizing solution before use [68].

When the membrane becomes scratched or dissolved so that the worn away membrane falls below the level of the electrode body, it can be restored by polishing. To polish the surface, it is necessary first to grind down the electrode body until the membrane becomes level with the body. This can be done [62] using a piece of North Metalite cloth (Grade 320). The electrode is held perpendicularly on a piece of the cloth placed on a flat glass or Perspex plate and moved back and forth with a stroking motion of about three inches. Meanwhile the electrode should be regularly inspected to make sure that no excess of either membrane or electrode body is removed. When the membrane and the electrode body are leveled, the surface of the membrane is polished using (John Oakey and Sons) emery polishing paper. Then a few drops of Orion Silicone oil (94-00-03) are placed on the polishing paper and the electrode is repolished to restore its glossy appearance.

Mechanical polishing of the electrode membrane to remove the salt deposits can be done using a toothbrush and toothpaste [23]. The membrane of the bromide electrode (AgBr membrane) after exposure to a high level of thiocyanate is cleaned by this method to remove silver thiocyanate precipitate.

Chemical treatment of the electrode membrane has also been utilized. The response time of the sodium electrode increases after prolonged use, probably due to the formation of a hydrated layer on the electrode surface. This layer can be removed [68, 69] by immersing the electrode for 30 s in an aqueous solution containing 50–100 mg of ammonium bifluoride per 100 ml.

Grease and waxy deposits on the fluoride electrode can be removed using "Parson's wax removal", "Fantastik" or a cleaning solution prepared by diluting 10 ml of concentrated aqueous ammonia and 10 ml of a detergent or wetting agent to 200 ml. The electrode is immersed in the cleaning solution and the temperature is raised to about 70°C for 3–5 min. Afterwards the electrode is transferred to distilled water previously heated to about 70°C, cooled to below 35°C, then immersed in 6 M hydrochloric acid for 30 s, and finally rinsed with distilled water [68].

Gas-sensing membrane probes, in which the sensor is a glass electrode, start to become sluggish after several months in use, because of deterioration in the response of the glass electrode. The response can be restored by alternate treatment of the glass electrode with dilute sodium hydroxide and hydrochloric acid solutions [51].

Most of the heterogeneous solid state and liquid ion-selective electrodes should be soaked in the standardizing solution before use. Such electrodes require daily restandardization.

D. Storage

After use, most of the solid state (except the cyanide electrode which should be stored dry) and the liquid ion-selective electrodes can be stored in air,

Table 1.3. Storage conditions of some electrodes

Storage condition	Electrodes
In air or immersed in a standard-izing solution	Cadmium (S), copper (S), fluoride (S), nitrate (L), potassium (L), silver/sulfide (S), calcium (L), chloride (L), divalent (L), perchlorate (L), boron tetrafluoride (L)
In air or immersed in distilled water	Chloride (S), bromide (S), iodide (S), thiocyan-ate (S)
In air or under dry conditions	Cyanide (S), lead (S)

S: solid state membrane. L: liquid membrane.

standardizing solution or distilled water. When the liquid ion-selective electrodes are stored in air, some of the ion exchanger seeps out around the edges of the membrane, and the electrode bottom should be cleaned to remove the ion exchanger material before use. The various commercially available electrodes can be stored according to the conditions given in Table 1.3.

E. Effect of operational conditions on the electrode performance

1. *Effect of Temperature*

Differentiation of the Nernst equation with respect to temperature gives three parameters [70]:

$$(dE/dT)_{th} = (dE^0/dT)_{th} + (0.19841/n) \log a$$
$$+ (0.1984\, T/n)(d \log a/dT) \qquad (22)$$

The first parameter (dE^0/dT) depends on the particular ion-selective electrode and on the reference electrode to be used. The second parameter $(0.19841/n) \log a$ is a temperature correction factor and can usually be performed by the manual or automatic temperature compensator in the pH-meter. The third parameter $(0.19841\, T/n)(d \log a/dT)$ is the solution temperature coefficient which depends on the effect of temperature on the activity of the solution.

Since the electrode potential and its slope are affected by the change in temperature of the sample (Table 1.4), the temperatures of both the standard solution and the sample should be the same at the time of measurement. At the 10^{-3} M level, a 1°C difference in temperature gives rise to a 2% error. The absolute potential of the reference electrode changes slowly with temperature because of the solution equilibrium on which the electrode depends.

Table 1.4. Effect of temperature on the slope of the electrode
potential

Temperature (°C)	Slope	
	Monovalent ions	Divalent ions
0	54·20	27·10
5	55·18	27·59
10	56·18	28·09
15	57·18	28·59
20	58·16	29·08
22	58·56	29·28
24	58·96	29·48
25	59·96	29·58
26	59·36	29·68
28	59·76	29·88
30	60·14	30·07
35	61·14	30·57
37	61·54	30·77
38	61·74	30·87
40	62·14	31·07
50	64·12	32·06

Although the ion-selective electrodes can be used at temperatures ranging from 0° up to 100°C provided that temperature equilibrium is attained, it is advisable to use the electrode around room temperature only. At temperatures substantially different from room temperature, equilibrium time may take up to an hour. However, when measurement at temperatures above 80° are required, it is advisable to use the electrode only intermittently, in view of the fact that the electrode life is shortened by continuous exposure to high temperature.

The effect of temperature on the performance of the gas-sensing membrane probes is more complex, because the different parts of the probes have different temperature coefficient and thermal capacities, and thus the overall temperature coefficient will apparently vary with time until the equilibrium is attained. The ammonia probe takes a much longer time to attain equilibrium than other probes, probably because an increase in the temperature of the sample induces osmosis. The apparent temperature coefficient of the ammonia probe increases to about 2 mV/°C, while the sulfur dioxide probe becomes virtually insensitive to temperature variations [51].

The effect of temperature on the shape of the e.m.f. curve of the carbon dioxide probe is remarkable. From the beginning it can be seen that the first movement is in the negative direction, but this is then reversed after a short time (approx. 3 min) and the potential becomes more positive than it was initially. After 10–15 min, the direction changes again and the potential slowly reaches a steady value [50].

In general, under laboratory conditions, care must be taken to ensure that standards and samples are kept at the same temperature and that the probe is not subjected to rapid temperature fluctuations or sunlight during measurements.

2. Effect of Radiation

The effect of β-radiation on the response of some electrodes has been a subject of investigation since 1949 [71]. It has been reported that silver chloride and calomel electrodes are stable over long periods of radiation while the glass and antimony electrodes are stable for shorter periods. The change of sensitivity gradient, asymmetrical potential and resistivity of lithium and glass electrodes in a gamma field, after doses up to $2 \cdot 5 \times 10^7$ rad, have also been studied [72]. These electrodes are unstable for pH measurements in media of high gamma activity.

Recently, the studies of the effect of radiation has been extended to cover some ion-selective membrane electrodes. The fluoride electrode [73] is affected by γ-radiation, but its usefulness can be maintained by periodic checks with standard fluoride solution. Doses in the order of 10^5 rad result in small parallel shifts which become non-linear above 10^6 rad. The linearity is restored with a slight displacement in potential from the initial when the internal reference solution of the electrode is replaced.

Liquid ion-selective electrodes (e.g. Orion nitrate Model 92-07) are stable to cumulative cobalt-60 γ-ray doses of 10^4 rad, although there is a small parallel shift in the negative direction of the calibration graph. Larger doses (up to 5×10^7 rad) cause non-linear responses [73]. The linearity is restored with a slight displacement in potential from the initial when the internal reference solution is replaced. Thus, with the liquid ion-selective electrodes, the major noticeable effect of radiation is on the reference solution, while the liquid ion exchanger changes to a dark color and the white electrode body changes to an amber color. These are the only noticeable signs of change after about a 5×10^7 rad exposure. In general, the commercially available electrodes can be used in a γ-radiation field of fairly high intensity (10^4 rad/min) with little abnormal behavior.

The effect of radiation has also been investigated on some heterogeneous poly(vinyl chloride) membrane electrodes containing Orion or Corning nitrate ion exchanger. These electrodes give essentially the same slope, constant, activity range and response time after exposure to a total γ-ray dose of $1 \cdot 6 \times 10^3$ rad over a period of 24 h as the non-irradiated electrode. Such electrodes can withstand short exposures to cobalt-60 γ-radiation without significant loss in performance [43, 74].

3. Effect of Organic Solvents

The use of solid state membrane electrodes in non-aqueous media has been investigated. With the fluoride electrode [75], the Nernst equation ($E = E_0 + RT/nFQ$) is obeyed in as many organic solvents as in water. Because the solubility of lanthanum fluoride decreases in the presence of ethanol, one would expect that the electrode can be used to cover large pF values in the presence of ethanol than in an aqueous medium. Titration of fluoride with lanthanum ion is further improved considerably by the addition of 60 to 70% by volume of ethanol. Titration of fluoride with lanthanum in 95% ethanol or acetone and processing the results using time response graph paper, which permits extrapolation of the electrode potential to infinite time [76], has been reported. The results (as well as measurements of conductivity and apparent pH) show that when buffer solutions containing ammonium chloride, ammonium acetate or ammonium formate are used, the accuracy is severely limited by the effect of hydrolysis and formation of undissociated hydrofluoric acid and/or lanthanum compounds. The range of concentration and the choice of usable polar solvents are greatly extended when an amino acid (preferably glycine) is used as a buffer solution at pH 6·35. Other media, each characterized by a low dielectric constant, can also be used (e.g. glycine-buffered mixture of ethanol-chloroform-5% of H_2O or ethanol-acetic anhydride-5% of H_2O) [76]. Interference from Cl^-, NO_3^-, or SO_4^{2-} decreases with an increase in the content of the organic component in aqueous–organic solvents [76a].

The dependence of the e.m.f. response on logarithmic fluoride activity in pure aqueous and in 50% (v/v) aqueous–organic solvents has been investigated. A slope of 59 mV per activity decade is reported down to 10^{-5} M fluoride solution in water as well as in 50% (v/v) aqueous ethyl alcohol, isopropyl alcohol, acetone, DMF and acetonitrile. With 10^{-6} M fluoride, the response becomes sluggish and a negative deviation by 20 mV from the anticipated Nernstian value is indicated in all these solvents [77]. Similar results are obtained in 60% and 80% (v/v) aqueous ethyl alcohol. In 50% aqueous dioxane, however, a typical Nernstian response is observed from 10^{-1} down to 10^{-6} M fluoride. In 80% dioxane, no improvement is noted and the electrode fails to respond ideally with 10^{-1} and 10^{-2} M fluoride solutions, probably due to insolubility of the fluoride. Therefore the use of 50% aqueous dioxane is recommended as an excellent background for fluoride ion determination.

It has been observed that the constant E_0 is more negative in the presence of organic solvents than in pure aqueous medium, and its value depends on both the nature of the solvent and the pH of the medium. The shift is mainly due to asymmetric potential across the membrane of the electrode because of the presence of different solvents on either side of it.

In 0·1 M sodium nitrate, the shift is in the order: DMF > dioxane > acetonitrile ≈ acetone > isopropyl alcohol > ethyl alcohol > water. This shift ranges from 40 mV in the case of ethanol up to 80 mV in the case of DMF. As the percentage of the solvent increases, the shift becomes 100 mV in ethanol and 150 mV in DMF. In 0·1 M hydrochloric acid, the shift is less pronounced and does not exceed 20 mV [77].

Study of the effect of alcohol homologues on the iodide ion-selective electrode shows that the dissolution of the electrode membrane decreases as the alcohol solvent changes from methanol to butanol [78]. Copper (II) ion-selective electrodes can be used in water–methanol and in pure methanol solutions [79]. In pure methanol, the electrode proved to be applicable for measuring copper (II) ions in the range 10^{-1} to 10^{-4} M. Different calibration curves are necessary for different aqueous–organic mixtures. However, this inconvenience can be eliminated by titrating the non-aqueous sample solution with an aqueous solution of ethylenediamine tetra-acetic acid disodium salt. Similarly, copper (II) in acetonitrile and acetone can be titrated. Also, copper (I) ions in aqueous as well as in non-aqueous media can be measured using the copper (I) ion-selective electrode [80]. In general, solid state electrodes with epoxy bodies must not be used in DMF, chloroform or similar polar solvents. The effect of organic solvents in the nitrate-selective electrode has been reported [81].

Heterogeneous solid state membrane electrodes have also been tested in non-aqueous solvents. Silicone rubber membrane electrodes do not respond in anhydrous solvents. However, they can be used in DMF, acetone and alcoholic solutions containing 10–40% water or acetonitrile solutions containing up to 40% water [82]. Membrane electrodes containing plastic matrices should not be used in organic solvents which might affect the membrane.

It should be noted that as a rule liquid membrane electrodes are used only in aqueous solutions. Organic solvents may dissolve or remove the liquid exchanger causing the electrode to fail.

Titrations in non-aqueous solvents using the ion-selective electrodes have been recommended when the solvent used decreases the solubility of the precipitate formed by titration or when the sample to be titrated is water-insoluble. For example, fluoride ion can be titrated with thorium in 80% ethanol [83] or 50% dioxane [77]. As the concentration of dioxane increases the potential break at the end point increases [84]. Titration of fluoride with lanthanum in 50% ethanol may also be used [85].

Titrations of sulfate and oxalate ions with lead (II) using the lead ion-selective electrode show that the reaction is non-stoichiometric due to the solubility of the precipitate. However, the use of p-dioxane as a medium for the titration decreases the solubility of lead sulfate and lead oxalate, thus permitting their quantitative analysis. Optimum results are obtained by titration in 60% [86–88] and 40% [89] dioxane for the sulfate and oxalate, respectively. It should be noted that the dioxane used must be fresh, because old dioxane usually contains peroxide which suppresses the potential of the lead electrode to such an extent that its use is impractical [90, 91].

Organic liquids can also be used as solvents for either the material to be titrated or the titrant. Thiol compounds are usually dissolved in p-dioxane ethanol or acetone before titration with mercuric ion using the bromide electrode [92]. p-Dioxane and toluene have been used as solvents for hydroxy polymers [93]. Preparation of mercuric perchlorate in 70–75% p-dioxane, ethanol or acetone before titration with mercuric ion using the bromide ion-selective electrode [94]. Iodide ions can also be titrated with silver nitrate in 90% C_1 to C_6-containing alcohols [95]. However, solvents of high vapor pressure are not recommended for preparing titrants, because their ease of volatility affects the strength of the titrant.

4. Effect of pH

Although many of the ion-selective electrodes can be used for direct measurements of ionic species over a wide pH range, hydrogen and hydroxyl ions interfere with some of these electrodes.

The influence of pH on the response of the cyanide electrode and ammonia probe can be predicted from a logarithmic diagram [31] of HCN/CN^-, and ammonium/ammonia systems respectively, or from the relation:

$$\overset{+}{N}H_4/NH_3 \text{ or } HCN/CN^- = 10^{9.2}/10^{pH}$$

It is apparent that the optimum pH range of both sensors is above 11. Below this pH, association of CN^- or ammonia with H_3O^+ occurs. However, these sensors can be used at pH values as low as 10, provided that the pH is carefully controlled.

The effect of pH on the performance of some solid state ion-selective electrodes has been studied. The fluoride electrode shows a negative shift in potential in basic media [96, 97]. The exact origin of this shift is complicated and explanation is still tentative. However, it seems that the interference of the hydroxide ion is of a simple additive type based on the relation:

$$E = E^0 - RT/F \ln \{[F^-] + k_s[OH^-]\} \tag{23}$$

The selectivity ratio k_s varies from zero in 10^{-1} M to greater than one in 10^{-5} M fluoride solution.

Although Frant and Ross [96] attributed the effect of the hydroxide ion to physical factors, based on its similarity to the fluoride ion in both charge and ionic radii, the possibility of a chemical reaction between the hydroxide ion and the lanthanum fluoride membrane may be considered.

The equilibrium of formation of lanthanum hydroxide complexes or precipitate, which might liberate free fluoride ion, obviously leads to a negative shift in the potential:

$$LaF_3(s) + 3OH^- \rightleftharpoons La(OH)_3(s) + 3F^-$$

Since the K'_{sp} of the lanthanum hydroxide [98] is $10^{-21.7}$ and the effective solubility product of lanthanum fluoride membrane is in the order of 10^{-22}, lanthanum fluoride dissolves to produce lanthanum hydroxide if

$$[OH^-]/[F^-] = (K'_{sp}/K_{sp})^{1/3} \qquad (24)$$

On the other hand, a positive shift in potential is observed in acidic solutions. Srinivasan and Rechnitz [40] studied the behavior of the fluoride electrode in acidic media to gain a clear understanding of its performance in terms of the complex fluoride species existing in such solutions. The average number of fluoride ions bound to hydrogen (the ligand number) in different acidic solutions was calculated from the equation of Bjerrum [99].

$$\bar{n} = [F^-]_{total} - [F^-]_{free}/[H^+]_{total} \qquad (25)$$

The data show that HF_2^- is the highest complex and there is no evidence for the presence of polynuclear species [100]. Thus, the species F^-, HF and HF_2^- are prevalent in the region of F^- and H^+ concentration. This work shows that the fluoride electrode selectively senses the free fluoride ions, even in acidic solution, and the response of the electrode towards HF_2^-, if present, is negligible [40]. In order to have 99% or more of the fluoride in the free form, the pH must be two units or more higher than the pK_a of the acid (i.e. pH above 5).

The sodium electrode cannot be used to measure low levels of sodium in acidic solutions and the pH of these solutions should be raised with basic reagents (e.g. tris-hydroxymethylamino methane, triethanolamine or sodium-free barium hydroxide) [31] before measurements. Formation of insoluble calcium, copper, cadmium and lead hydroxides limits the pH range over which these ions can be measured with their respective electrodes. The hydroxides precipitate at a higher pH in dilute rather than in concentrated solutions. The effects of changes in ionic strength on ion-selective electrodes for Na^+, K^+ and Ca^{2+} determinations in serum have been found to be relatively minor [100a].

The liquid ion-exchange chloride electrode can be used for the measurement of the chloride ion at pH 2 to 10–11. The upper pH limit depends on the chloride level. At pH 10 (10^{-4} M OH^-) a solution containing 10^{-1} M chloride exhibits no error, a 10^{-2} M solution shows a 1% error, and a 10^{-3} M solution shows a 10% error. For work below pH 2 and between pH 11–12, the solid state chloride electrode is preferable since it has a wider usable pH range [33].

Calcium liquid ion-exchanger electrodes can be used over a pH range of 5·5 to 11 with little error due to changes in pH. The electrode is believed to give correct readings for calcium activity above pH 11, but difficulties are experienced in the interpretation of the data due to formation of calcium hydroxide. In the acid regions (below pH 4), the electrode responds to hydrogen ions and no longer measures calcium ions [34]. This is probably because the hydrogen ion contributes to the charge transport process across the membrane, producing a positive increase in the electrode potential. With a further decrease in the pH, the contribution of the calcium ion to charge transport in the membrane is overwhelmed by the hydrogen ion, and the electrode responds in a Nernstian manner as a pH electrode. If the calcium electrode is immersed for a long time in acidic solution, the calcium phosphate ester salt in the membrane phase changes to the acidic form. The exchanger can be recovered again by placing the electrode in a calcium solution of high pH for a long time.

Potentiometric titration of ionic species should be conducted under a suitable pH to assure quantitative reaction and to avoid interferences. The optimum pH for titration of some common ions using the ion-selective electrodes is given in Table 1.5.

Table 1.5. Optimum pH for titration of some ionic species using the ion-selective electrodes

Ion titrated	Titrant	Optimum pH	Electrode
Arsenate (+ excess La)	Fluoride	8·65	Fluoride
Cadmium	EDTA	10	Cadmium
Calcium	EDTA	11	Calcium
Copper	EDTA	10–11	Copper
Cyanide	Silver	7	Cyanide
Fluoride	Thorium or Lanthanum	5–7	Fluoride
Halides	Silver	6–7	Halide
Lead	EDTA	4–7	Lead
Nitrate	Diphenyl thallium (III) sulfate	2–4	Nitrate
Oxalate	Lead	3·5–9·5	Lead
Phosphate (+ excess La)	Fluoride	4·8	Fluoride
	Lead	8·2–8·8	Lead
Sulfide	Silver	10	Silver/sulfide
Thiocyanate	Silver	7	Thiocyanate or Silver

III. EQUIPMENT AND OPERATION TECHNIQUES

A. Ordinary laboratory operations

The use of ion-selective electrodes for the analysis of ionic species in the laboratory does not require expensive equipment. Two basic units are needed besides the ion-selective electrode: a reference electrode and a pH/mV meter. However, concentration measurements can be read directly on specific ion meters specially designed for this purpose (e.g. Orion specific ion meter models 407, 407A, 701, 901 and Beckman Century SS-1 meter). For direct measurements, a pH/mV meter calibrated to ±0·1 mV or less should be used; for potentiometric titrations, meters calibrated to ±10 mV may be sufficient.

Anion-reversible reference electrodes are generally more useful than cation-responsive reference electrodes, presumably due to their stability. A saturated or 3·8 to 0·1 M potassium chloride calomel reference electrode and an additional salt bridge, if halide ions are incompatible with the test solution, may be used. The conventional fiber and frit types are not recommended because the low rate of leakage usually exhibits drifting and unstable junction potential [101, 102].

The use of a second ion-selective electrode as a reference electrode, with an intermediate salt bridge has been suggested for fluoride measurements [103]. The sample is titrated until the cell e.m.f. becomes zero. This technique, known as null point potentiometry, is described in detail in Chapter 3, Section IVC.

Combination electrodes are available for fluoride, chloride, sodium and redox determinations. These electrodes contain both the sensing element and the reference electrode, and allow direct measurements on very small samples, such as those on surfaces, filter paper or confined spot test paper.

The choice of the reference electrode is important (see Table 1.6), since a drastic change in the slope of the calibration graphs can be related to some malfunction of the reference electrode.

Analysis of an ionic species in solution is based on the measurement of the developed e.m.f. by an electrochemical cell composed of the ion-selective electrode and the reference electrode or by using the combined electrode or the gas-sensing membrane probe. The sample solution is placed in a beaker, microdish or on the membrane surface of an inverted electrode. The potential developed in the solution which is related to the activity of the measured ion can be monitored by using one of the measurement techniques described in detail in Chapter 3.

Interference due to the presence of undesirable ions in the background or matrix effect can be eliminated by addition of a few milliliters of a suitable ionic strength adjustor (ISA) and/or pH adjustor to both the standard and sample solution before measurement (Table 1.7). These

Table 1.6. Some recommended reference electrodes and their filling solutions

Ion-selective	Reference electrode used	References
Bromide	—Double junction with 4 M KCl saturated with silver ion in the outer compartment —Double junction with 1 M KNO_3	
Cadmium	—Single junction —Double junction with 10% KNO_3	104
Calcium	—Single junction with 4 M KCl saturated with silver ion —Saturated calomel electrode	105–110
Chloride	—Double junction with 1 M KCl and sucrose —Double junction with 1 M KNO_3 —Double junction with 0·1 M KCl	111 112
Copper	—Single junction with 10% KNO_3 —Double junction with 4 M KCl for solutions of pH > 1 —Double junction with HNO_3 for solution of pH < 1	112 113 114
Cyanide	—Single junction with 10% KNO_3	
Fluoride	—Single junction with 1 M KNO_3	
Fluoroborate	—Double junction with 0·1 M NH_4F	115, 116
Lead	—Double junction with 50% methanol saturated with KCl —Double junction with 1 M $NaNO_3$	112
Nitrate	—Double junction with 0·01 M KF or 0·1 M KF —Double junction with 0·1 M KCl —Double junction with Hg/Hg_2SO_4 in 1 M Na_2SO_4	117–120 121–125 126
Potassium	—Single junction with lithium chloro acetate/lithium chloride —Single junction with 0·06 M NaCl	127
Sodium	—Single junction with lithium chloro acetate/lithium chloride	
Sulfide	—Double junction with 10% KNO_3	128, 129
Thiocyanate	—Double junction with 1 M KNO_3	

solutions are composed of either concentrated salt solutions or acid, base and buffer solutions depending on the appropriate pH range.

Measurement of ionic species in gaseous mixtures (e.g. samples of air containing HF, SO_2 and NO_2) can be conducted by passing the air or gas stream through a suitable entrapping material [134]. Filter papers

Table 1.7. Some ionic strength adjustors

Ion measured	Reagent	Reagent/sample ratio (v/v)
Ammonia, cyanide	10 M NaOH	1 : 100
Bromide, chloride, iodide, copper, cadmium, lead, silver, thiocyanate	5 M NaNO$_3$	2 : 100
Calcium, water-hardness	4 M KCl	2 : 100
Fluoroborate, perchlorate, nitrate	10 M NH$_4$F	1 : 100
Fluoride	TISAB^a	1 : 10
Potassium	8 M NaCl	1 : 100
Sodium	10 M NH$_4$Cl + 1 M NH$_4$OH	1 : 100
Sulfide	SAOB^b	1 : 1

^{a} TISAB—"Total Ionic Strength Adjustor Buffer" is prepared by one of the following procedures: (i) Dissolve 57 ml of acetic acid, 58 g of NaCl, 0·3 g sodium acetate in 500 ml of water and the solution is titrated to pH 5 with 5 M NaOH and diluted to 1 liter [130]. (ii) Dissolve 170 g of NaNO$_3$ and 68 g sodium acetate trihydrate in 1 liter of water [131]. (iii) Dissolve 17·65 g of diamino cyclohexanetetra acetic acid (DCTA) in 500 ml of water and add 40% NaOH solution dropwise until the salt dissolves. Then add 300 g of sodium citrate dihydrate and 60 g NaCl, dilute to 1 liter with water and adjust the pH to 6 with HCl [132].

^{b} SAOB—"Sulfide Anti-Oxidant Buffer" is prepared by dissolving 80 g NaOH in about 500 ml water, then 320 g of sodium salicylate is slowly added. After complete dissolution, 70 g of ascorbic acid is added and the solution is completed to 1 liter [133].

impregnated with 5% potassium carbonate [134a], 10% suspension of calcium oxide [135] or sodium formate [136] can be used and the trapped species can be extracted with a suitable solvent and measured with a suitable ion-selective electrode. The amount of the ion in the solution, as given by the electrode, may be converted to ion per unit size of paper and in turn to ion concentration in air or gas stream.

B. Flow-through, continuous on-line and automatic operations

The use of ion-selective electrodes for continuous measurements in flowing systems and their incorporation in totally automated monitoring systems have been established. This has wide application in heavy industry, process control, pollution control, pharmaceutical and medicinal analyses. Sample handling equipments, as well as computer interfacing systems have been developed.

Electrodes used for continuous measurement should have quick response and not be influenced by rate of flow of the solution to be measured. The solid state electrodes can be used satisfactorily because the response of most of these electrodes is related to a fast surface ion-exchange equilibrium reaction. However, some electrodes do not maintain a constant potential during prolonged measurements, and all liquid membrane electrodes cannot be employed at high rates of flow.

In continuous monitoring, the signal due to the concentration must be evaluated either by calibration with standards or by using the standard addition technique [137]. The measuring system should be frequently checked because of the possible drift of the electrode function and consequent change of the electrode calibration graph.

1. Flow-through Operations

Various measuring cells are used in flow-through analysis [137a,b]. These cells are characterized by: (i) having a small size to permit the analysis of small volumes; (ii) the ease of cleaning and arrangement of the detector units; and (iii) the ease of replacement of the electrodes during operations. Cells made of a transparent material can be used to enable the visual following of the measurements.

Flow-through caps (Fig. 1.5) can be used in conjunction with many electrodes. The flow-through systems have several advantages over static type electrodes—rapid equilibrium time (usually 30–60 s), stability, applicability to small sample sizes (~0.2 ml), and possible measurability in anaerobic conditions. These types of electrodes are particularly useful in measuring ionized species in biological fluids and in continuous on-line-flow. The sample can flow through these electrodes using a gear-driven syringe pump (Fig. 1.6).

A flow-through cell with two compartments separated by an ion-selective membrane may be used. A stream containing the standard or reference flows directly over the surface of the membrane in one compartment, and the test sample is allowed to flow over the other surface of the membrane [138] (Fig. 1.7).

A flow-through cell in which the carrier solution flows tangentially over the membrane surface of the ion-selective electrode and from there to the reservoir where the terminal part of the reference electrode is submerged, has been reported [139]. Since a constant and small level of the liquid is maintained in the reservoir, a pumping rate of a few milliliters of the solution per minute is sufficient. This device can be used for the flow injection analysis of ionic species (Fig. 1.8).

Flow-through cells have also been designed to be used for on-line monitoring under controlled conditions of pH and temperature. Water-jacketed electrodes and Perspex mixing chambers of the type shown in Fig. 1.9 can be used [140]. A flow-through electrode unit incorporating an inverted cone solution cavity can also be used [141] to permit: (i) regulated stirring of the test solution, (ii) constant solution temperature, (iii) steady liquid-level recirculation, and (iv) housing for the necessary instrumentation. The design (Fig. 1.10) behaves as an ideal well-stirred vessel and

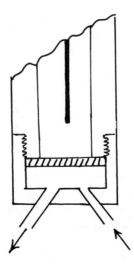

FIG. 1.5. Flow-through cap.

provides both solution depth and sensitivity to small changes in liquid level while allowing for free circulation at the base of the cavity.

2. *Continuous On-line Operation*

Application of ion-selective electrodes and gas sensing probes to continuous on-line monitoring can be achieved by placing the electrode system in the sample stream provided that the solution to be measured is not corrosive,

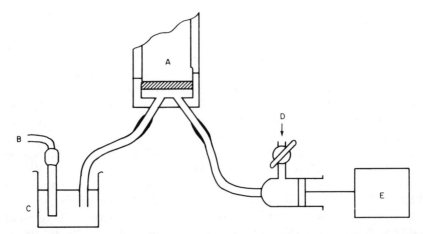

FIG. 1.6. Flow-through electrode in continuous on-line flow. A, Flow-through ion-selective electrode; B, reference electrode; C, saturated potassium chloride solution; D, sample inlet; E, pump.

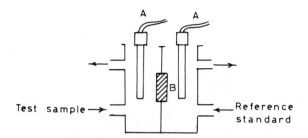

FIG. 1.7. Flow-through cell. A, reference electrode; B, ion-selective membrane.

contains no suspended particles or interfering ions, and its composition, temperature and pH do not vary over a period of time [141a–c]. The sample stream continuously flows past the electrode and the outputs of the electrodes and the temperature compensator are fed into a converter which permits continuous recording and control of the process stream [142].

Multiple ion-selective electrodes in a flow-through cell may be used for measuring several constituents. For example, if the cyanide electrode is used for monitoring cyanide ions in neutral or acidic solutions, the electrode readings can only be interpreted with a knowledge of the pH, and therefore a pH and a cyanide electrode are used simultaneously. Applications of the electrodes for monitoring gases in a stream can be performed by scrubbing the gas constituents with a suitable reagent and measuring the quantity of the gas and reagent with flow transmitters [136].

When the electrode system cannot be directly immersed in the sample stream due to the presence of corrosive substances or interfering ions, a sample is taken from the main pipe through a narrow tube and treated with a suitable reagent, using proportionating pumps before

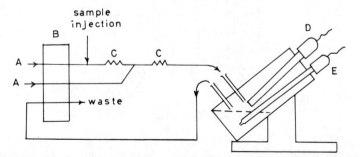

FIG. 1.8. Confluence manifold with a flow-through cell for flow injection analysis of the nitrates (according to Hansen, Ghose and Ruzicka, *Analyst* **102**, 705 (1977)). A, buffer solution; B, pump; C, mixers; D, nitrate ion-selective electrode; E, reference electrode. (Courtesy of The Chemical Society of London.)

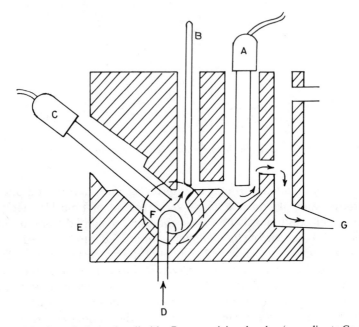

FIG. 1.9. Flow-through electrode cell with a Perspex mixing chamber (according to Covington, *Chemistry in Britain* **5**, 388 (1969)). A, reference electrode; B, thermometer; C, ion-selective electrode; D, monitor water in; F, water-jacketed thermostatically regulated coil; G, waste out. (Courtesy of The Chemical Society of London.)

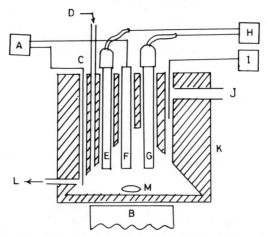

FIG. 1.10. Thermostatically regulated flow-through electrode cell (according to Forney and McCoy, *Analyst* **100**, 157 (1975)). A, temperature controller; B, stirring unit; C, temperature probe; D, solution in; E, ion-selective electrode; F, heater; G, reference electrode; H, pH/mV meter; I, solution level probe; J, overflow; K, sensing cell; L, solution out; M, bar. (Courtesy of The Chemical Society of London.)

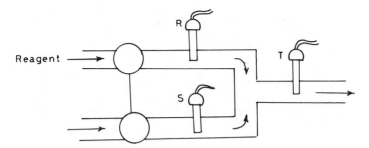

FIG. 1.11. Measurement in a flowing system with three ion-selective electrodes (according to *Orion Newsletter* **2**, 1 (1971)). (Courtesy of Orion Research Inc.)

measurements [143]. The reagent prevents interference from the presence of oxidizing, reducing, complexing, basic and acidic substances.

To avoid the problems associated with the use of reference electrodes, such as variation of liquid junction potential, clogging and the need of periodical refilling, a reference electrode with a slow flow or immobilized poly(vinyl chloride) reference electrodes may be used. On the other hand, when a pair of identical specific ion electrodes is used, the electrode drift is cancelled. The ion-selective electrode used in place of the reference electrode should be sensitive to another ion whose level is held constant. Consequently, a "tag ion" is added to the reagent. For example, when the fluoride ion is the species to be determined and a TISAB solution, which contains a constant level of sodium is used, a sodium ion-selective electrode is utilized as a reference.

Measurements in flow-systems can also be performed [143] using three identical electrodes as depicted in Fig. 1.11: S in the sample stream, R in the reagent stream and T in the mixed stream. The potential differences $(E_R - E_S)$, $(E_R - E_T)$ and $(E_T - E_S)$ give the activity of the sample by direct measurements, the concentration of the sample by the analate addition method, and the concentration of the sample by the known addition method, respectively.

3. *Automatic Operations*

Automatic analyzers with an electrode module containing either a flow-through specific ion-electrode and reference electrode or gas-sensing membrane probe have been developed. Commercial types of these analyzers are now available (e.g. Orion series 3300) by which runs with about 60 samples per hour can be conducted with internal restandardization on every analysis for precise and reproducible work. In these systems, the sample and reagent are mixed using a peristaltic pump and a dynamic mixer, then passed through a thermostatically regulated electrode head (Fig. 1.12) attached to a high impedance recording system.

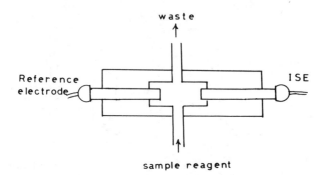

FIG. 1.12. Thermostatically regulated electrode head.

Analyzers for continuous measurement (e.g. Orion Series 1000) have also been developed for the analysis of many ions (e.g. Model 1132 for sulfide, 1229 for copper, 1107 for nitrate, 1125 for chlorine, 1109 for fluoride and 1206 for cyanide). These systems (Fig. 1.13) are built to perform in the rigorous environment of the industrial plant. The electrode assembly is electronically controlled by a thermostat to allow actual measurements to be made under laboratory conditions. Clean samples for

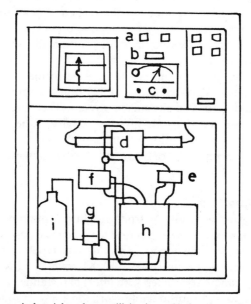

FIG. 1.13. Continuous industrial analyzer utilizing ion-selective electrodes (Orion series 1000 monitors). A, alarm indicator light; B, automatic standardization range; C, dual set-point indicating meter; D, electrode head; E, mixer; F, heater; G, valve; H, pump; I, standardizing solution. (Courtesy of Orion Research Inc.)

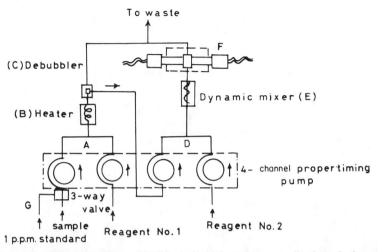

FIG. 1.14. Cyanide monitor (Orion 1206) for industrial continuous applications. A, the sample is mixed with EDTA (reagent No. 1); B, the stream is heated to 80°C to release metal bound cyanide; C, air bubbles are removed; D, a basic reagent (reagent No. 2) is added to the stream; E, the reagent and sample are mixed; F, the stream is passed to the electrode chamber; G, a standard solution is introduced into the system automatically through a 3-way valve to test the system. (Courtesy of Orion Research Inc.)

the monitor are provided by a by-pass filtration system at a minimum supply of 2000 to 4000 ml/min at 25 psi. The results can be displayed on a strip chart recorder with a dual set-point indicating meter. Such systems give results accurate to ±10% and have a limit of detection of about 0·1 p.p.m. (Fig. 1.14).

An electrode analyzer for simultaneous measurement of the pH, CO_2, Na, K, total calcium and ionized calcium in blood and urine for in-flight analysis in space, has been developed by Orion Research Inc. through a contract with NASA [144]. The system developed was called SPACE-STAT (stat from the Latin *statim*; immediately). This system permits the analysis of less than 1 ml of a sample without any effect due to gravity or pressure that might be exerted in space and with minimum reagent and waste solutions. It consists of four electrode modules containing seven electrodes—sodium and potassium electrodes in one module, pH and ionized calcium in a second module, total calcium and chloride in a third, and carbon dioxide in the fourth. Each module contains a common reference electrode and liquid junction so that the samples are simultaneously measured for two parameters by both electrodes (Fig. 1.15). The chloride electrode used is the solid state $AgCl/Ag_2S$ crystal membrane; calcium and potassium electrodes have a liquid membrane and gelled internal filling solution; sodium and pH electrodes have a glass membrane and carbon dioxide gas-sensing membrane probe. The analytical package contains two

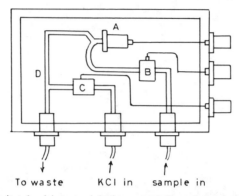

To waste   KCl in   sample in

FIG. 1.15. Chloride/total calcium module for automatic analysis. A, calcium ion-selective electrode; B, chloride ion-selective electrode; C, reference electrode; D, liquid junction. (Courtesy of Orion Research Inc.)

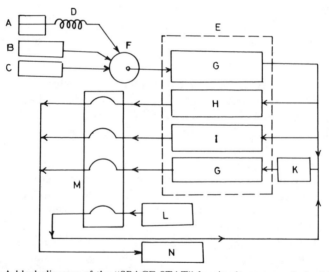

FIG. 1.16. A block diagram of the "SPACE STAT" for simultaneous analysis of CO_2, Na, K, Ca, Cl and pH (according to *Orion Newsletter* **6**, 9 (1974)). A, sample injection port; B, C, standard solutions; D, holding loop; E, electrode bank; F, multiport switching valve; G, CO_2/reference electrode system; H, Na/K/reference electrode system; I, ionized calcium/pH/reference electrode system; J, total calcium/chloride/reference electrode system; K, mixer; L, acetic acid reagent; M, peristaltic pump; N, waste container.

The sample enters the injection port and is held in the holding loop until it comes to temperature equilibrium with the system. The multiport switching valve determines the system measurement sequence. Standard B enters the CO_2 electrode module; then the sample is split into three portions for measurement by the remaining modules. Before the sample enters the total Ca/Cl module, it is mixed with acetic acid solution to displace protein-bound calcium. After analysis, all spent solutions are pumped into the waste container. (Courtesy of Orion Research Inc.)

standardizing solutions to calibrate the seven electrodes. The samples and standardizing solutions are moved through the electrodes in sequence, controlled by the positions of the multiport valve. The timing of the positions of the valves is determined by the computer. A simplified block diagram of the analytical package is shown in Fig. 1.16. The system is supplied with three programs: (i) a control program to determine the sequence and timing of the valves, peristaltic pump, mixer and thermostat; (ii) a computation program to calculate the unknown concentration and print out the results on teletype; and (iii) a service program to isolate problems, to test the various subsystems and to facilitate trouble shooting in the system.

Following the same principle, automatic analyzers for ionized calcium (Orion SS-20) and serum sodium/potassium (Orion SS-30) have been developed, and are available in the market. The samples are injected in these systems as in case of gas/liquid chromatography, and the results are read from the digital display in terms of concentration. A minimum sample size of 1 ml can be automatically introduced from vials in the sample tray or, a collecting syringe is used to deliver a 500 μl-sample through the injection port. The results can be read directly so that 1–3 min are only required per sample including the internal standardization (Fig. 1.17).

It should be noted that in all the automatic or continuous monitoring systems, glass or homogeneous and heterogeneous solid state membrane electrodes should be used. Liquid membrane electrodes are not recommended due to their fast exhaustion in flowing or continuous systems. The

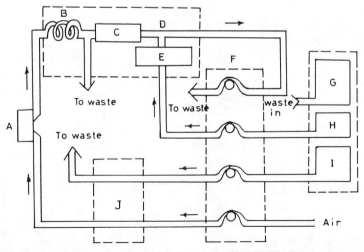

FIG. 1.17. A block diagram of ionized calcium analyzer (SPACE STAT Orion-20) according to Orion instruction manual. A, injection port; B, holding coil; C, calcium sensor; D, liquid junction; E, reference electrode; F, peristaltic pump; G, waste; H, KCl; I, standard; J, valve. (Courtesy of Orion Research Inc.)

calcium electrodes used in the aforementioned automatic systems are of the heterogeneous solid state type (i.e. a calcium ion exchanger embedded in PVC). When gas-sensing membrane probes are used in continuous systems, the electrode should not be immersed in the solution but fixed in a position to sense the measurable species in the gaseous phase. Such electrode systems should be renewed after six months.

Automatic on-line potentiometric titration using a minicomputer titrator has been described [145]. The input voltage signals from the ion-selective electrodes are transferred *via* an analogue scanner to a digital voltmeter where they are digitized before being transferred to the minicomputer *via* a digital scanner. An integrated automated titration system based on the use of a PDP-8/I minicomputer has been proved to be useful in a variety of analytical applications using ion-selective electrodes [146].

Continuous titration can be used for the analysis of some ions. For example, calcium can be determined using cadmium-EDTA as the indicator and the system is shown in Fig. 1.18. By means of a proportioning pump, cadmium-EDTA and the sample solution are mixed with sodium EDTA, *via* a variable speed pump. The potential of the cadmium/reference electrode system is used to control the variable speed pump in order to maintain the electrode potential exerted at the titration end point. The calcium level in the sample stream is proportional to the flow rate of the variable speed pump, which is indicated by flow meters [143]. A similar method has been used for continuous monitoring of the nickel level in a high-speed nickel plating bath using the copper electrode [112].

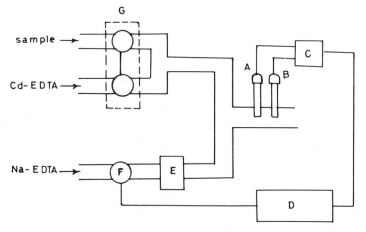

FIG. 1.18. Continuous titration of calcium using Cd-EDTA and calcium ion-selective electrode (according to *Orion Newsletter* **2**, 23 (1970)). A, cadmium ion-selective electrode; B, reference electrode; C, electrode amplifier; D, pump rate control; E, flow-meter; F, variable rate pump; G, proportioning pump. (Courtesy of Orion Research Inc.)

C. Voltammetric operations

The silicone rubber-based graphite electrodes can be used as voltammetric indicator electrodes. They are used in conjunction with either calomel or silver/silver chloride reference electrodes. The voltammetric signal is flow-rate dependent, because it is a result of the transport process. These electrodes are used successfully for the measurement of the flow-rates in the range of 0·1 mm–50 cm/s [147]. In streaming solutions, the electrode reaction is controlled by convective diffusion. Correlations have been given for the relationships between the voltammetric current and the flow-rate (Fig. 1.19).

Voltammetric measurements [148, 149] can be performed by placing the graphite electrodes in a thin tube through which the supporting electrolyte flows at a constant rate. The test solution is then injected into the tube just before the electrodes and the time function of the transient signal is determined. After mixing, the test solution enters the flow-through cell containing the indicator and reference electrodes. The change in current intensity is recorded at a constant voltage corresponding to the limiting current range of the component to be measured. A peak-type signal is obtained and the effect of the flow-rate and the concentration of the material tested on the area under the peak is established. Thus, the amount of the electro-active material injected can be determined by integrating the current or potential with respect to time, which corresponds to the area under the peak. The linear relationship between the area under the peak and the amount of the injected electro-active material at a constant flow-rate is then constructed. By this technique, it is possible to determine many electro-active materials. The error is within ±0·1% if the temperature is controlled, and if the electrodes have a well defined surface.

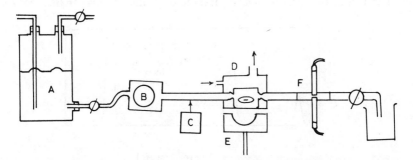

FIG. 1.19. Continuous voltammetric analysis using the injection technique and ion-selective electrodes (according to Feher, Nagy, Toth and Pungor, *Analyst* **99**, 699 (1974)). A, container; B, peristaltic pump; C, injection unit; D, thermostat; E, stirrer; F, detector ion-selective electrode cell. (Courtesy of The Chemical Society of London.)

Using a suitable electro-active reagent, electro-inactive samples can also be analyzed using this technique. Such a determination can be carried out in two ways, depending on the concentration level of the electro-inactive sample. At lower concentration ranges, the electro-inactive sample is allowed to flow continuously, but at the higher concentration ranges the measurement is more convenient if the electro-inactive sample is injected.

The silicone rubber based graphite electrodes have been used for *in vivo* measurements. Distribution of drugs in living organisms can be followed [150]. The change in the current intensity with time can be seen as an effect of some drugs (Fig. 1.20). The effect of aminopyrine when injected into the rear limb of a narcotized test animal can be followed. The current recorded varies at a constant average value, which corresponds to the actual drug level in the blood. This variation is due to changes in the pulse rate of the blood according to the heart beats.

A special voltammetric cell has been fabricated in which a turbulent flow is achieved by pumping the solution to be measured with a powerful pump through a tube with a narrow jet against the sensing surface of the silicone rubber-based graphite indicator electrode. The enhanced convection effect obtained resulted in a higher sensitivity of the voltammetric measurements. This cell is used as a chromatovoltammetric detector operating at a constant potential. When this cell is connected to a suitable chromatographic column, separation and detection of various samples (e.g. adenine, guanine, xanthine and hypoxanthine) is accomplished. The lower limit of detection of these compounds is 2×10^{-10} mol. The sensitivity of the detector is about the same as that of ultra-violet detectors, but it has the advantages of selective detection [151].

Carbon paste electrodes impregnated with ceresin wax and silicone oil, in conjunction with calomel electrodes, have been used for voltammetric analysis of vitamins A, B_6, C and E [152–155]. A conventional three-electrode polarograph can be used and the voltammograms can be recorded.

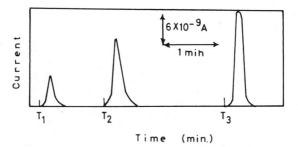

Time   (min.)

Fig. 1.20. Current–time curves recorded in the artery after administration of aminopyrine and voltammetric measurement. Amounts of drug injected are 0·4, 0·8 and 1·2 mg/kg. U = +0·8 V, T = time of sample injection (according to Feher, Nagy, Toth and Pungor, *Analyst* **99**, 699 (1974)). (Courtesy of The Chemical Society of London.)

D. Chromatographic operations

One of the most important applications of the ion-selective electrodes is as detectors in certain chromatographic operations. This is based on the principle that ions regarded as interfering with a particular electrode can be readily detected by the electrode after chromatographic separation. Thus, liquid membrane electrodes are very useful for the determination of mixtures of ions after liquid chromatographic separation. These electrodes are sensitive to nanomole quantities of inorganic and organic anions and can be extended to cationic species as well.

The nitrate ion-selective electrode has been used [123] as a detector for the determination of nitrate–nitrite mixtures and phthalate isomers (Fig. 1.21). Samples of 1–$100\,\mu$l containing 40–100 nmol of each component

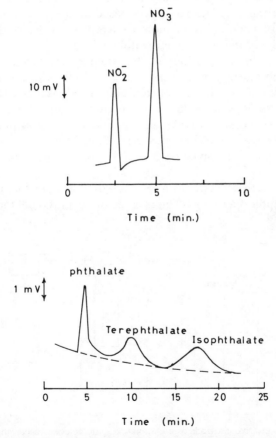

FIG. 1.21. Separation of nitrate–nitrite mixture and phthalate isomers using the nitrate ion-selective electrode and ion-exchange liquid chromatography (according to Schultz and Mathis, *Anal. Chem.* **46**, 2253 (1974)). (Courtesy of the American Chemical Society.)

are injected in the chromatographic column and the ions eluted under a pressure of 500 psi are detected with the nitrate electrode. The electrode is fitted with a flow-through cap with a small cavity ($\sim 5 \; \mu$l)—drilled in the inside of the bottom of the cap—to serve as a detection chamber. The column is connected to the electrode with polyethylene tubing and the ion-selective electrode and the reference electrode are immersed in a beaker containing the eluent solution (Fig. 1.22). The electrode response is measured with a pH-meter and strip chart recorder and the peak height is based on the relation:

$$E = -(E_s - E_b) = 0 \cdot 0591 \log \left[\frac{a_{NO_3} + K}{K} \right] \qquad (26)$$

where E_s is the maximum signal with the sample and E_b is the base line signal. Logarithmic calibration curves based on this equation can be used. Thus, plots of peak areas *vs* mole of samples can be constructed. The reproducibility of this technique is within $\pm 5\%$.

By changing the composition of the liquid ion-exchange solute used as a liquid membrane, it is possible to obtain detectors which respond to many other organic and inorganic cations and anions and to enhance the selectivity for a group of ions. One limitation of the liquid-membrane electrode detector, however, is that highly selective anions such as nitrate and perchlorate ions and some halides must be excluded from the eluent. Solutions containing sulfate, phosphate and borate can be used due to their low selectivity.

The ion-selective electrodes have been used as detectors for gas–liquid chromatographic operations. This is based on separation of the components

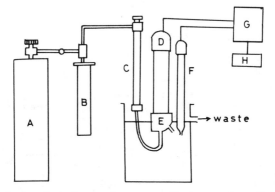

FIG. 1.22. Ion-selective electrode as a chromatographic detector (according to Schultz and Mathis, *Anal. Chem.* **46**, 2253 (1974)). A, nitrogen tank; B, solvent reservoir; C, ion-exchange column; D, liquid membrane ion-selective electrode; E, flow-through cap; F, reference electrode; G, pH/mV meter; H, recorder. (Courtesy of the American Chemical Society.)

on a gas–liquid chromatographic column using hydrogen as a carrier gas and the effluent is passed through a platinum tube at 1000°C where the separated components undergo hydrogenolysis. The gaseous decomposition products are then conducted into an absorption tube containing a suitable absorbent, and the solution emerging from the tube is passed into a micro-cell equipped with a suitable electrode which responds to the ionic species formed (Fig. 1.23). Changes in concentration in the solution are detected by the changes in the electrode potential, which is then fed to an antilogarithmic converter circuit, and signals directly proportional to the ion concentration are recorded [156, 157].

This technique has been satisfactorily utilized for the separation and determination of some fluorine compounds admixed with sulfur, nitrogen and halogen containing compounds. The detection limit is very low (e.g. 5×10^{-11} mol for fluorobenzene) and the method is superior to that based on the use of flame ionization or thermal conductivity detectors [156, 157].

Mixtures of thiols can be resolved and determined by gas–liquid chromatography using the silver/sulfide ion-selective electrode as a detector [157]. The components are eluted from a 2 m column packed with 5% of silicone DC on chromosorb W (80 to 100 mesh) and are passed through an absorption tube. The effluent from the tube is then passed into a

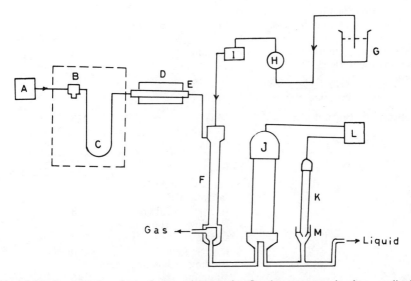

FIG. 1.23. Ion-selective electrode as a detector for fluorine compounds after gas–liquid chromatographic separation (according to Kojima, Ichise and Seo, *Talanta* **19**, 539 (1972)). A, hydrogen tank; B, sample injection port; C, chromatographic column; D, electric furnace; E, pyrolysis tube; F, glass absorption tube; G, absorption solution; H, pump; I, buffer; J, fluoride ion-selective electrode; K, reference electrode; L, potassium chloride-agar bridge; M, recording system. (Courtesy of Pergamon Press.)

micro-cell equipped with a silver/sulfide electrode and the change in silver ion concentration is monitored potentiometrically. An hydrogenolysis reactor can also be inserted after a column packed with 25% of tritolyl phosphate on Shimalite W (80–100 mesh), and the sulfide ion formed is absorbed in silver ion solution and the change in the concentration can be monitored.

E. Biochemical and biomedical operations

Since ion-selective electrodes actually measure ion activity rather than concentration, they can be successfully used for biological measurements because biological phenomena are functions of ionic activities rather than concentrations. Study of the protein binding of ions and the measurement of a particular ion activity in the presence of other ions *in vivo* and *in vitro* can be conducted in biological fluids, tissues and organs (Table 1.8). The commercially available electrodes can be used in some purposes but specially designed macro-, micro- and ultramicro-electrodes have been developed.

Table 1.8. Analysis of some biological, biochemical and biomedical substances by the ion-selective electrodes

Ion measured and electrode used	Biological material	References
Ammonia by ammonia gas sensing membrane probe	Plasma	211
	Feces	212
Bromide by bromide electrode	Plasma	209
Calcium by calcium electrode	Milk	213, 214
	Blood	195, 215–224
	Serum	196, 219, 225–252
	Plasma	236, 253–256
	Binding submaxillary saliva in CF	257
	In hyperparathyroidism	258
	Binding by chondroitin sulfate	259
	Gastric juice	260
	Binding to bovine serum albumin	246
	Albumin aggregate	261
	During exchange transfusion	216
	Biological fluids	262
	Kidney fluids	222

Table 1.8 (cont.)

Ion measured and electrode used	Biological material	References
Chloride by chloride electrode	Cystic Fibrosis (CF)	207, 208, 263, 264
	Milk	265, 266
	Serum	267
	Urine	267
Fluoride by fluoride electrode	Saliva	200, 268–273
	Enamel	274–278
	Plaque and tooth	204, 206, 273, 276, 279–283
	Bone	284–286
	Urine	200, 273, 287–296
	Serum	200–203, 297–300
	Feces	205
	Fish and proteins	301, 302
	Milk	200, 303
Magnesium by water hardness electrode	Binding with nucleic acid	210
Nitrate by nitrate electrode	Microbial media	304
	Vegetation, fruit	305
Potassium by potassium electrode	Biological fluids	174, 306
	Blood serum	182, 185, 224, 307
Sodium by sodium electrode	Serum	172
	Blood	170, 308
	Bile	173
	Muscles	309
	Intracellular	310
	Brain	170
	Clinical medicine	202, 311
Urea by urea electrode	Biological fluids	212

It should be noted that the protein, which is one of the components in most biological materials, is the main poisoning substance of many electrodes. However, glass membrane electrodes are, in general, less susceptible to protein poisoning and any protein coat formed on the electrode surface can be easily removed by rinsing the probe promptly with a buffer solution.

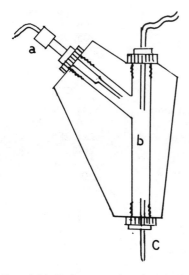

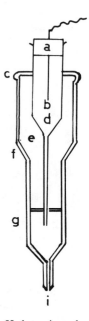

FIG. 1.24. Reference micropipette electrode. a, calomel electrode; b, 3 M KCl
coloured solution; c, reference micropipette. (Courtesy of the National Bureau
of Standard Publications.)

FIG. 1.25. pH glass micro-electrode. a, oil;
b, Ag/AgCl electrode; c, cement; d, 0·01 M
HCl; e, oil; f, pH glass; g, polystyrene; i,
10–20 μm sensitive tip. (Courtesy of the
National Bureau of Standard Publications.)

The glass electrodes have wide applications in biological study not only
for pH measurements but also for the analysis of many cations. The main
precautions are to avoid dryness and overhydration of the glass membrane,
and the successive use of the electrode in solutions of widely varying
compositions which may cause hysteresis.

The reference electrode used should be selected so that a minimum
diffusion of the filling solution takes place. Micropipette reference electrodes filled with potassium chloride solution, which acts as a salt bridge
[158], are usually used. To prevent flow of potassium chloride caused by
hydrostatic force, the electrode is used either in the horizontal position or
by using the potassium chloride filling solution in the form of gel in 1–2%
agar. An internal capillary micro-electrode dipped in the salt bridge solution
[159] or, a reference micropipette electrode [160] which is mounted into
a holder and allows the injection of potassium chloride solution through
its tip (Fig. 1.24), can also be used.

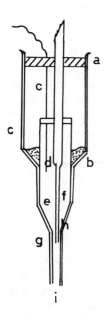

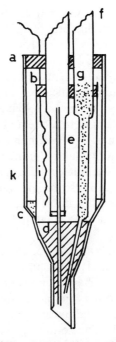

FIG. 1.26. Glass ultramicro internal capillary electrode. a, b, wax; c, Ag/AgCl electrode; d, h, glass-to-glass seal; e, reference solution; f, capillary glass electrode; g, micropipette; i, 8–12 μm tip. (Courtesy of the National Bureau of Standard Publications.)

FIG. 1.27. Combination pH/reference ultramicro internal capillary electrode. a, b, c, d, wax; e, buffer of pH 7; f, calomel; g, 3 M KCl; h, 2% agar in 3 M KCl; i, Ag/AgCl electrode; j, pH glass; k, protecting glass. (Courtesy of the National Bureau of Standard Publications.)

1. Measurements of pH

Micro and ultramicro glass electrodes of the spear-shaped bulb or the internal capillary type are usually used for *in situ* pH measurements. The electrodes may have a flat, convex or concave surface depending on the curvature of the tissue under study. The spear-shaped electrodes are suitable for deep penetration and the flat membrane type is used for extracellular measurements. The response time of these electrodes is short so that changes due to metabolic alterations may be picked up promptly. For an *in vivo* study, a very small probe should be used, since tissue damage may increase the local acidity.

A micropipette-type pH glass electrode (Corning 0150) which is insulated by mineral oil internally and polystyrene paint externally (Fig. 1.25), has been used for *in vivo* measurements of the intraluminal pH in the proximal tubule of animal kidneys [160, 161]. Internal capillary micro-electrodes (Fig. 1.26) can be used for *in vivo* pH measurements of tubular fluids [162].

A combined internal capillary pH ultramicro-electrode with a reference half-cell in the form of a single unit [163], permits measurement on solution of volumes as low as 0·05 μl and is applicable to the *in vivo* measurements [162] (Fig. 1.27).

pH measurements of biological fluids especially in the kidney *in vivo* are of medical significance, since the kidney plays an important role in the acid–base balance of living organisms and excretes the excess acids of metabolism to regulate and maintain a constant pH in the blood and other body fluids. Measurements of the pH in single renal nephrons [160, 161] and in intracellular regions have been reported by several groups [164–169]. The extracellular pH changes during contraction and relaxation of skeletal muscles can also be measured. *In situ* pH measurements in the brain of animals under different metabolic conditions can similarly be performed by introducing a small pH probe into the brain.

2. *Measurements of Sodium and Potassium*

Sodium and potassium are two prevalent cations in many biological materials. They are present in serum, bile, blood, brain and muscle tissues and in other biological fluids in various proportions. Glass electrodes sensitive to sodium and potassium ions have been widely utilized in biomedical work. The sodium glass membrane electrodes have selectivity in the order: $Ag^+ > H^+ > Na^+ \gg K^+$, $Li^+ \gg Ca^{++}$ and the potassium glass membrane electrodes have selectivity in the order: $H^+ > K^+ > Na^+ \gg Ca^{++}$. However, in a neutral, high sodium and low potassium biological system like plasma where the sodium/potassium ratio is about 35–40, the sodium electrode essentially responds to sodium ions, whereas the use of the potassium electrode necessitates prior sodium measurement to obtain the potassium concentration. This is possible by using a sodium/glass electrode couple. All the electrodes used for sodium measurements are of the glass membrane type whereas those used for potassium measurements are either of the glass or liquid membrane type.

Sodium glass membrane electrodes may be used to measure the sodium ion activity in animal brains *in vivo* [170] and *in situ* [171]. Human serum [172] and bile [173] can also be analyzed. Sodium in circulating blood, glomerular fluids and proximal tubular fluids of the Necturus kidney [174] has been measured. The sodium electrode has been used in the diagnosis of cystic fibrosis of the pancreas which is characterized by an elevated concentration of sodium and chloride ions in sweat [175].

Potassium glass membrane electrodes have been utilized to monitor potassium ion activity in animal brains *in vivo* [170, 171] and in many other biological systems [170, 176–179]. A liquid membrane potassium electrode based on the use of potassium tetra (*p*-chlorophenyl) borate dissolved in 3-nitro-*o*-xylene [180, 181] as a liquid exchanger has been used for potassium ion measurements in serum, living tissues and the intracellular space of individual cells [182]. Heterogeneous solid-state potassium ion-selective electrodes prepared from valinomycin in 4–8% poly(vinyl chloride) and 10–18% dipentyl phthalate have been used for potassium measurement in serum [183].

A micropipette containing a liquid ion exchanger in its tip has been designed for potassium measurements (Fig. 1.28). The terminal of the pipette tip (~200 μm) is coated with silicone to overcome the hydrophilic nature of the clean glass [184]. A double-barrel electrode (Fig. 1.29) based on the work of Coombs *et al.* [185] can be used for ion activity measurements in situations where it is necessary to avoid the electrical activity of either single cells or whole tissues [184–186]. The double-barrel potassium electrode has been used for potassium measurements in the parietal cortex of animals [187, 188]. A side pore type of ion-selective electrode (Fig. 1.30) has been developed for measurement of extracellular potassium concentration in muscles. This design does not cut the muscle fibers and prevents undesirable movement of the ion exchanger into the muscles during measurements [189].

Intracellular analyses of sodium and potassium ions in the protoplasm of single nerve fibers, muscle and alga Nitella can be made using the sodium and potassium micro-electrodes [179, 190]. It should be noted that measurements made in small cells are not simple because it is difficult to see the tip of the cell. However, the abrupt shift in the potential of the electrode according to the equation shown below (equation 27) when the

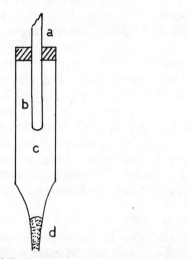

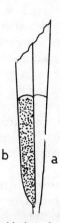

FIG. 1.28. Walker's potassium electrode. a, mineral oil; b, Ag/AgCl electrode; c, 0·5 M KCl; d, liquid ion-exchanger.

FIG. 1.29. Double-barrel potassium electrode. a, NaCl; b, organic liquid exchanger.

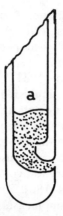

FIG. 1.30. Side-pore potassium electrode. a, liquid ion-exchanger.

electrode is slowly moved towards the cell while the reference electrode remains outside can be taken as an indicator for cell entry.

$$E = \Delta O_M + RT/F \ln \frac{(a_K)_e + K_{K,Na}^{pot}(a_{Na})_e}{(a_K)_i + K_{K,Na}^{pot}(a_{Na})_i} \qquad (27)$$

where ΔO_M denotes the cell membrane potential and the indices e and i characterize the quantities pertaining to the extracellular and intracellular space respectively [184].

Extracellular measurements can be done on non-flowing and flowing systems [191, 192]. Measurements of sodium and potassium ions *in situ* in cerebrospinal fluids and on the brain surface, skeletal muscles, aqueous humor, the lumen of the stomach and gall bladder have been reported. Sodium and potassium are measured in luminal fluids of single proximal tubules *in situ* using sodium, pH or antimony micro-electrodes by localizing the electrode within the lumen of the kidney's tubule [160–162, 174, 193]. Extracellular measurements of ionic activity in circulating fluids can be made by maintaining a constant flow rate to avoid fluctuation of potential readings. Flow-through electrodes of sodium and potassium glasses have been used to measure the cationic activity in the flowing arterial blood of animals [192, 193]. Simultaneous analysis of sodium and potassium can be automatically performed using the Orion SS-30 analyzer.

3. *Measurement of Calcium*

Calcium is found in biological materials in the form of protein-bound calcium, diffusible calcium complexes and ionized calcium. Fortunately, ionized calcium is the physiologically active species and a large number of its important physiological processes include bone formation, muscle contraction, nerve conduction, resorption, cardiac conduction and contraction of the cerebral function, intestinal secretion and absorption, renal tubular function, blood coagulation, membrane and capillary permeability, enzyme function and hormonal release from various endocrine glands. These are now known to be critically dependent on the calcium ion activity. Consequently, the invention of the calcium ion-selective electrode has provided a powerful new tool for biological and biomedical studies.

However, the commercial static version of calcium ion-selective electrodes do not work satisfactorily in the presence of proteins. This problem can be circumvented by either using the flow-through electrode system or separation of ionized calcium from protein. A high pressure dialysis cell for ultrafiltration or a protective cellophane dialysis membrane can be used for protein separation [182].

Calcium in sera has been determined after protein separation using a viscose dialysis tubing with an approximately mean pore diameter of 25 Å. After formation of 2–3 ml of the ultrafiltrate, each sample is removed anaerobically from the cell and immediately introduced into a Plexiglass chamber previously aerated with 5% carbon dioxide for electrode analysis of the ionized calcium [194]. In general, the measurement of calcium should be done only on anaerobically prepared serum without treatment with anticoagulants. Untreated whole blood clots in the flow-through electrode, while ionized calcium levels in heparinized whole blood and plasma are changed by the addition of anticoagulants. Centrifugation in the collection syringe and the use of weakly dissociated acid heparine salts as anticoagulants have been suggested [195].

The calcium electrode has been used for the study of the effect of parathyroid extract and thyrocalcitonin on ionized calcium levels in subjects with parathyroid diseases [196, 197]. Ionized calcium has also been determined in subjects with cirrhosis and certain types of cancer. A calcium electrode may also be used to follow the variations in ionized calcium in patients being treated for calcium deficiency using artificial kidney machines [198]. Since the binding of calcium by protein increases with the increase of the pH, ionized calcium levels can be measured in newborn babies with acidosis (lower blood pH) caused by respiratory diseases [199].

4. Measurement of Fluoride
The quantitative measurement of fluoride in biological materials is difficult using the conventional analytical methods. However, fluoride in urine, enamel, plaque, dentine, teeth, saliva, serum, bone, feces, proteins, milk, soft tissues and biological fluids can be measured easily and rapidly using the fluoride ion-selective electrode.

Fluoride is present in plasma in non-bound, ionic and bound forms associated with albumins. The former fraction can be directly measured by the electrode after protein separation by diffusion from hydrochloric acid at room temperature or perchloric acid at 60°C [200]. Fluoride in serum and protein-bound fluoride can be determined after ashing with a mixture of magnesium chloride and sodium carbonate [201–203]. Bone, enamel, dentine, plaque and teeth are decomposed by pyrohydrolysis at 700–1000°C in the presence of vanadium pentoxide or tungsten oxide [204]. Excreted fluoride in urine, sweat and feces and in milk, saliva and hair can be measured after ashing at 550°C [205], or directly in unashed samples or by suspending the materials in perchloric acid [206].

5. *Miscellaneous Biological Measurements*
One of the most important medical applications of the chloride ion-selective electrode is the rapid diagnosis of cystic fibrosis, a disease appearing at birth and characterized by an increase in the concentration of sodium and chloride ions in sweat on the skin. The method is based on the use of moistened pads of electrolyte and pilocarpin to induce the formation of sweat in a small area of the skin in a short time by applying an electronically controlled current to carry the pilocarpin into the skin. The chloride electrode is then placed on this area and the chloride level is read directly on the meter. This method is safe, reliable and it can be applied to large-scale screening programs in newborns, infants and children [207, 208].

Bromide [209], magnesium [210], ammonia [211] and urea [212] electrodes have also been utilized for the analysis of biological materials including milk, plasma, microbial media, nucleic acid binding, urine, feces and serum (see Table 1.8).

REFERENCES

1. R. A. Durst (ed.), *Ion-Selective Electrodes*. U.S. Government Printing Office, Washington (1969).
2. E. Pungor and I. Buzas (eds), *Ion-Selective Electrodes*. Elsevier, Amsterdam (1978).
3. R. G. Bates, Electrode Potentials. *In* I. M. Kolthoff and P. J. Elving (eds), *Treatise on Analytical Chemistry*, 2nd Ed., Vol. 1, Part 1. Wiley, New York (1978).
3a. E. Guggenheim, *J. Am. Chem. Soc.* **52**, 1315 (1930).
3b. M. Planck, *Ann. Physik. Chem.* **39**, 161 (1890).
3c. W. Nernst, *Z. Physik Chem.* **2**, 613 (1888).
4. F. Donnan, *Z. Elektrochemie* **17**, 572 (1911).
5. T. Teorell, *Proc. Soc. Exp. Biol. Med.* **33**, 282 (1935).
6. K. Meyer and J. Sievers, *Helv. Chim. Acta* **19**, 649 (1936).
7. E. Pungor and K. Toth, *Analyst* **95**, 625 (1970). E. Linder, K. Toth and E. Pungor, *Anal. Chem.* **48**, 1071 (1976); **50**, 1627 (1978).
8. D. Mackay and P. Meares, *Kolloid-Z.* **171**, 139 (1960).

9. R. Buck, *Anal. Chem.* **40**, 1432 (1968); **40**, 1439 (1968).
10. F. Conti and G. Eisenman, *Biophys. J.* **5**, 247 (1965); **5**, 511 (1965).
11. G. Karrenman and G. Eisenman, *Bull. Math. Biophys.* **24**, 413 (1962).
12. M. Shults, Dokl, *An SSSR* **194**, 337 (1970).
13. B. Nikolsky, *Zh. Fiz. Khim.* **10**, 495 (1937).
14. G. Eisenman, *Anal. Chem.* **40**, 310 (1968).
15. J. Sandblom, *J. Phys. Chem.* **73**, 249 (1969).
16. J. Sandblom, G. Eisenman and J. Walker, Jr, *J. Phys. Chem.* **71**, 3862 (1967); **71**, 3871 (1967).
17. J. Walker, G. Eisenman and J. Sandblom, *J. Phys. Chem.* **72**, 978 (1968).
18. Z. Stefanac and W. Simon, *Microchem. J.* **12**, 125 (1967).
19. G. Eisenman, S. Ciani and G. Szabo, *Fed. Proc.* **27**, 1289 (1968).
20. C. Pedersen, *J. Amer. Chem. Soc.* **89**, 7017 (1967).
21. P. Mueller, D. Rudin, H. Tien and W. Wescott, *Circulation* **26**, 1167 (1963).
21a. J. G. Schinder, H. K. Duerr, W. Riemann, H. E. Braun, and V. Kellner, *Biomed. Tech.* **23**(3), 45 (1978).
22. E. Pungor and K. Toth, *Anal. Chim. Acta* **47**, 291 (1969).
23. Orion Research Inc., *Newsletter* **1**, 29 (1969).
24. K. Toth and E. Pungor, Proc. Intern. Measurement Confederation Symposium Electrochem. Sensors, Veszprem, Hungary, p. 35 (1968).
25. Orion Research Inc., Instruction Manual, Cyanide Electrode (Model 94-06) (1972).
26. G. Moody and J. Thomas, *Lab. Practice* **20**, 307 (1971).
27. K. Srinivasan and G. Rechnitz, *Anal. Chem.* **41**, 1203 (1969).
28. K. Cammann and G. Rechnitz, *Anal. Chem.* **48**, 856 (1976).
28a. E. W. Baumann, *Anal. Chim. Acta* **99**, 247 (1978).
29. P. Debye and E. Hückel, *Phys. Z.* **24**, 185 (1923).
30. J. Hem, U.S. Geol. Surv. Cir. No. 1535-C (1961).
31. G. Rechnitz, M. Kresez, S. Zamochnick, T. Hseu and Z. Lin, *Anal. Chem.* **38**, 973, 1786 (1966); **40**, 696, 1054 (1968); **41**, 111 (1969).
32. J. Koryta, *Ion-Selective Electrodes.* Cambridge University Press, Cambridge (1975).
33. G. J. Heijne and E. E. van der Linden, *Anal. Chim. Acta* **96**, 13 (1978).
34. J. W. Bixler, R. Nee and S. P. Perone, *Anal. Chim. Acta* **99**, 225 (1978).
35. I. Sekerka and J. F. Lechner, *Anal. Chim. Acta* **93**, 139 (1977).
35a. J. C. Westall, F. M. M. Morel and D. N. Hume, *Anal. Chem.* **51**, 1792 (1979).
36. Orion Instruction Manual, Fluoride Electrode (1975).
37. P. Evans, G. Moody and J. Thomas, *Lab. Practice* **20**, 644 (1971).
37a. R. C. Hawkings, L. P. V. Corriveau, S. A. Kushneriuk and P. Y. Wong, *Anal. Chim. Acta* **102**, 61 (1978).
38. V. Ebock and C. Neiser, *Z. Chem.* **18**, 343 (1978).
38a. R. Durst and J. Taylor, *Anal. Chem.* **39**, 1483 (1967).
39. J. Lingane, *Anal. Chem.* **39**, 881 (1967).
40. K. Srinivasan and G. Rechnitz, *Anal. Chem.* **40**, 509 (1968).
41. R. Durst, *Anal. Chem.* **41**, 2089 (1969).
42. G. Moody, R. Oke, J. Thomas and J. Davies, *Analyst* **95**, 910 (1970); **97**, 87 (1972).
43. H. James, G. Carmack and H. Freiser, *Anal. Chem.* **44**, 856 (1972).
44. B. Kneebone and H. Freiser, *Anal. Chem.* **45**, 449 (1973).
45. J. Gulens and B. Ikeda, *Anal. Chem.* **50**, 782 (1978).
46. S. Lal and G. Christian, *Anal. Lett.* **3**, 11 (1970).
47. Orion Instruction Manual of Potassium Electrode (1973).
48. Orion Instruction Manual of Calcium Electrode (1972).
48a. U. Gruenke, P. Hartmann and J. Siemroth, *Hermsdorfer Tech. Mitt.* **17**, 1547 (1977).

49. I. Krull, C. Mask and R. Cosgrove, *Anal. Lett.* **3**, 43 (1970).
50. D. Midgley, *Analyst* **100**, 386 (1975).
51. P. Bailey and M. Riley, *Analyst* **100**, 145 (1975).
52. G. Papariello, A. Mukherji and C. Shearer, *Anal. Chem.* **45**, 790 (1973).
53. G. Guilbault and G. Nagy, *Anal. Chem.* **45**, 417 (1973).
54. G. Rechnitz and G. Kugler, *Anal. Chem.* **39**, 1682 (1967).
55. *Orion Newsletter* **3**, 11 (1971).
56. G. Perley, *Anal. Chem.* **21**, 559 (1949).
57. G. Rechnitz and H. Hameka, *Z. Anal. Chem.* **39**, 1683 (1967).
58. G. Johansson and K. Norberg, *J. Electroanal. Chem.* **18**, 239 (1968).
59. R. Buck, *J. Electroanal. Chem.* **18**, 363 (1968); **18**, 381 (1968).
60. R. Buck and I. Krull, *J. Electroanal. Chem.* **18**, 387 (1968).
61. *Orion Newsletter* **3**, 8 (1971).
62. *Orion Newsletter* **1**, 27 (1969).
63. *Orion Newsletter* **2**, 36 (1970).
64. *Orion Newsletter* **6**, 1 (1974).
65. C. Coetzee and H. Freiser, *Anal. Chem.* **40**, 2071 (1968).
66. E. Bauman, L. Goodson, G. Guilbault and D. Kramer, *Anal. Chem.* **37**, 1378 (1965).
67. Orion Instruction Manual of Lead Electrode (1972).
68. *Orion Newsletter* **5**, 14 (1973).
69. Orion Instruction Manual of Sodium Electrode (1973).
70. A. deBethune, T. Light and N. Swendeman, *J. Electrochem. Soc.* **106**, 616 (1959).
71. E. Kinderman and W. Carson, U.S. At. Energy Comm. Rep. TID-280 (1949).
72. N. Fedotov, *At. Energy* **8**, 262 (1960).
73. H. Kubota, *Anal. Chem.* **42**, 1593 (1970).
74. G. Moody and J. Thomas, *Proc. Soc. Anal. Chem.* **8**, 84 (1971).
75. J. Lingane, *Anal. Chem.* **40**, 935 (1968).
76. E. Heckel and P. March, *Anal. Chem.* **44**, 2347 (1972).
76a. L. I. Manakova, N. V. Bausova, V. E. Moiseev, V. G. Bamburov and A. P. Sivoplyas, *Zh. Anal. Khim.* **33**, 1517 (1978).
77. S. S. M. Hassan, *Mikrochim. Acta* 1974, 889.
78. L. Pataki, K. Harka, J. Havas and G. Keomaly, *Radiochem. Radional. Lett.* **27**, 385 (1976).
79. G. Rechnitz and N. Kenny, *Anal. Lett.* **2**, 395 (1969).
80. L. Heerman and G. Rechnitz, *Anal. Chem.* **44**, 1655 (1972).
81. A. Hulanicki, M. Maj-Zurawska and R. Lewandowski, *Anal. Chim. Acta* **98**, 151 (1978).
82. N. Kazaryan and E. Pungor, *Anal. Chim. Acta* **60**, 193 (1972); **51**, 213 (1970); *Megy. Kem. Foly.* **77**, 186 (1971).
83. T. Light and R. Mannion, *Anal. Chem.* **41**, 107 (1969).
84. C. Harzdorf, *Z. Anal. Chem.* **245**, 67 (1969).
85. W. Selig, *Mikrochim. Acta* 1970, 337.
86. J. Ross, Jr and M. Frant, *Anal. Chem.* **41**, 967 (1969).
87. W. Selig, *Mikrochim. Acta* 1970, 168.
88. R. Heistand and C. Blake, *Mikrochim. Acta* 1972, 212.
89. W. Selig, *Microchem. J.* **15**, 452 (1970).
90. Orion Research Inc., Anal. Methods Guide, 2nd Ed., Jan. 1972, p. 13.
91. W. Selig and A. Salomon, *Mikrochim. Acta* 1974, 663.
92. W. Selig, *Mikrochim. Acta* 1973, 453.
93. W. Selig, *Mikrochim. Acta* 1972, 612.
94. W. Selig and G. Crossman, *Z. Anal. Chem.* **253**, 279 (1971).

95. N. Kazaryan, L. Bykova and N. Chernova, *Zh. analit. Khim.* **31**, 334 (1976).
96. M. Frant and J. Ross, *Science* **154**, 1553 (1966).
97. R. Buck and S. Strecker, *Z. Anal. Chem.* **235**, 322 (1968).
98. L. Sillen and A. Martell, *Stability Constants of Metal-ion Complexes.* Special publication No. 17, the Chemical Society, London (1964).
99. J. Bjerrum, *Metal Ammine Formation in Aqueous Solution.* P. Haase and Son Pub., Copenhagen (1957).
100. L. Sillen, *Rec. Trav. Chim.* **75**, 705 (1956).
100a. M. S. Mohan and R. G. Bates, Natl. Bur. Stand. Spec. Publ. No. 450, p. 293 (1977).
101. D. Ives and G. Janz, *Reference Electrodes.* Academic Press, New York and London (1961).
102. A. Covington in R. Durst, *Ion Selective Electrodes*, p. 107. National Bureau of Standards, Special publication 314 (1969).
103. R. Durst, *Anal. Chem.* **40**, 931 (1968).
104. G. Bronow, T. Ilus and G. Miksche, *Acta Chemica Scand.* **26**, 1117 (1972).
105. E. Woolson, J. Axley and P. Kearney, *Soil Sci.* **109**, 279 (1970).
106. K. Cheng, J. Hung and D. Prager, *Mikrochem. J.* **18**, 256 (1973).
107. W. Wood, *J. Res. U.S. Geol. Survey* **1**, 237 (1973).
108. A. Hulanicki and M. Trojanowicz, *Anal. Chem.* **18**, 235 (1973).
109. T. Hadjiioannou and D. Papastathopoulos, *Talanta* **17**, 399 (1970).
110. S. Tackett, *Anal. Chem.* **41**, 1703 (1969).
111. L. Kopito and H. Schwachman, *Pediatrics* **43**, 794 (1969).
112. M. Frant, *Plating* July 1971, 686.
113. M. Smith and S. Manahan, *Anal. Chem.* **45**, 836 (1973).
114. Orion Instruction Manual, Cupric Ion Electrode (Model 94-29) (1973).
115. R. Carlson and J. Paul, *Soil Sci.* **108**, 266 (1969).
116. R. Carlson and J. Paul, *Anal. Chem.* **40**, 1292 (1968).
117. B. McCaslin, W. Franklin and M. Dillon, *J. of the A.S.S.B.T.* **16**, 64 (1970).
118. T. Henscheid, K. Schoenrock and P. Berger, *J. of the A.S.S.B.T.* **16**, 482 (1971).
119. E. Shaw and P. Wiley, *Calif. Agr.* **5**, 11 (1969).
120. T. Sommerfeldt, R. Milne and G. Kozub, *Comm. Soil Sci. Plant Anal.* **2**, 415 (1971).
121. P. Chalk and D. Keeney, *Nature* **229**, 42 (1971).
122. J. Bremner, L. Bundy and A. Agarwal, *Anal. Lett.* **1**, 837 (1968).
123. F. Schultz and D. Mathis, *Anal. Chem.* **46**, 2253 (1974).
124. J. Ridden, R. Barefoot and J. Roy, *Anal. Chem.* **43**, 1109 (1971).
125. D. Langmuir and R. Jacobson, *Envir. Sci. and Tech.* **4**, 834 (1970).
126. D. Capon, *Wat. treat. Exam.* **24**, 333 (1975).
127. G. Rechnitz and M. Mohan, *Science* **168**, 1460 (1970).
128. E. Baumann, *Anal. Chem.* **46**, 1345 (1974).
129. J. Barica, *J. Fis. Res. Bd Can.* **30**, 1589 (1973).
130. M. Frant and J. Ross, *Anal. Chem.* **40**, 1169 (1968).
131. A. Liberti and M. Mascini, *Anal. Chem.* **41**, 676 (1969).
132. M. Peters and D. Ladd, *Talanta* **18**, 655 (1971).
133. E. Green and D. Schnitker, *Marine Chem.* **2**, 111 (1974).
134. T. S. Ma and V. Horak, *Microscale Manipulations in Chemistry.* Wiley, New York, p. 345 (1976).
134a. K. Mukai and H. Ishida, Paper presented at the American Institute of Mining, Metallurgy, and Petroleum Engineers, Denver, Colorado, Feb. 16–18 (1970); *Orion Newsletter* **2**, 12 (1970).
135. D. Levaggi, W. Oyung and M. Feldstein, Paper presented at the 10th State Health Department Methods Conference, San Francisco, California, Feb. 19–21 (1969).
136. L. Elfers and C. Decker, *Anal. Chem.* **40**, 1658 (1968).

137. M. Brand and G. Rechnitz, *Anal. Chem.* **42**, 1172 (1970).
137a. W. E. van der Linden and R. Oostervink, *Anal. Chim. Acta* **101**, 419 (1978).
137b. A. Malinaukas and J. Kulis, *Anal. Chim. Acta* **98**, 31 (1978).
138. T. S. Light in Ref. 1, p. 349.
139. E. Hansen, A. Ghose and J. Ruzicka, *Analyst* **102**, 705 (1977).
140. A. Covington, *Chemistry in Britain* **5**, 389 (1969).
141. L. Forney and J. McCoy, *Analyst* **100**, 157 (1975).
141a. D. C. Cowell, *Med. Lab. Sci.* **35**, 265 (1978).
141b. A. D. Hirst, P. Gay, P. Richardson and P. J. N. Howorth, Natl. Bur. Stand. Spec. Publ. No. 450 (1977), p. 311.
141c. L. Liberti and A. Pinto, *Anal. Chem.* **49**, 2377 (1977).
142. R. Babcock and K. Johnson, *J. Amer. Water Works Assoc.* **60**, 953 (1968).
143. *Orion Research Newsletter* **2**, 1 (1971).
144. *Orion Newsletter* **6**, 7 (1974).
145. T. Anfalt and D. Jagner, *Anal. Chim. Acta* **57**, 177 (1971).
146. W. Selig, J. Frazer and A. Kray, *Mikrochim. Acta* 1975, 675.
147. E. Pungor, Z. Feher and G. Nagy, *Anal. Chim. Acta* **51**, 417 (1970).
148. E. Pungor, Z. Feher and G. Nagy, *Hung. Sci. Instr.* **18**, 37 (1970).
149. G. Nagy, Z. Feher and E. Pungor, *Anal. Chim. Acta* **52**, 47 (1970).
150. Z. Feher, G. Nagy and E. Pungor, *Hung. Sci. Instum.* **26**, 15 (1973).
151. Z. Feher, G. Nagy, K. Toth and E. Pungor, *Analyst* **99**, 699 (1974).
152. P. Soderhjelm and J. Lindquist, *Analyst* **100**, 349 (1975).
153. S. Atuma and J. Lindquist, *Analyst* **98**, 886 (1973).
154. J. Lindquist, *Analyst* **100**, 339 (1975).
155. S. Atuma, K. Lundström and J. Lindquist, *Analyst* **100**, 827 (1975).
156. T. Kojima, M. Ichise and Y. Seo, *Talanta* **19**, 539 (1972); *Bunseki Kagaku* **20**, 20 (1971).
157. T. Kojima, Y. Seo and J. Sato, *Bunseki Kagaku* **24**, 772 (1975).
158. G. Ling and R. Gerard, *J. Cell. Comp. Physiol.* **34**, 382 (1949).
159. O. Siggaard-Andersen, *Scand. J. Clin. Lab. Invest.* (Suppl.) **15**, 70 (1963).
160. R. Khuri, *Rev. Sci. Instrum.* **39**, 730 (1968).
161. F. Rector, N. Carter and D. Seldin, *J. Clin. Invest.* **44**, 278 (1965).
162. R. Khuri and S. Agulian, *Proc. Intl. Union Physiol. Sciences* **7**, 236 (1968).
163. R. Khuri, S. Agulian, H. Oelert and R. Harik, *Pflugers Archiv.* **294**, 291 (1967).
164. P. Caldwell, *J. Physiol* **126**, 169 (1954).
165. P. Caldwell, *J. Physiol.* **142**, 22 (1958).
166. P. Kostyuk and Z. Sorokina, *In* A. Kleinzeller and A. Kotyk (eds), *Membrane Transport and Metabolism*. Academic Press, New York and London, pp. 193–203 (1961).
167. M. Lavallee, *Circ. Res.* **15**, 185 (1964).
168. N. Carter, *Clin. Res.* **9**, 177 (1961).
169. N. Carter, F. Rector, D. Campion and D. Seldin, *J. Clin. Invest.* **46**, 920 (1967).
170. F. Gotoh, Y. Tazaki, K. Hamaguchi and J. Meyer, *J. Neurochem.* **9**, 81 (1962).
171. H. Portnoy, L. Thomas and E. Gurdjian, *Arch. Neurol.* **8**, 597 (1963).
172. E. Moore and D. Wilson, *J. Clin. Invest.* **42**, 293 (1963).
173. E. Moore and J. Dietschy, *Amer. J. Physiol.* **206**, 1111 (1964).
174. R. Khuri, D. Goldstein, D. Maude, C. Edmonds and A. Solomon, *Amer. J. Physiol.* **204**, 743 (1963).
175. P. Sekelj and R. Goldbloom, *In* G. Eisenman (ed.) *Glass Electrodes for Hydrogen and other Cations.* New York, Marcel Dekker, pp. 540–553 (1968).
176. G. Ling, *J. Gen. Physiol.* **43**, 149 (1960).
177. A. Lev, *Nature* **201**, 1132 (1964).

178. J. Hinke and S. McLaughlin, *Can. J. Physiol. Pharm.* **45**, 655 (1967).
179. J. Hinke, *J. Physiol.* **156**, 314 (1961).
180. G. Baum and W. Wise, Ger. Patent U.S. No. 2024636 (1970).
181. G. Baum and M. Lynn, *Anal. Chim. Acta* **65**, 393 (1973).
182. M. Wise, M. Kurey and G. Baum, *Clin. Chem.* **16**, 103 (1970).
183. M. Smith, M. Genshaw and J. Greyson, *Anal. Chem.* **45**, 1782 (1973).
184. J. Walker, *Anal. Chem.* **43**, 89A (1971).
185. J. Coombs, J. Eccles and P. Fatt, *J. Physiol.* (London) **130**, 291 (1955).
186. F. Vyskocil and N. Kriz, *Pflügerr Arch.* **337**, 265 (1972).
187. D. Prince, H. Lux and E. Neher, *Brain Res.* **50**, 489 (1973).
188. F. Vyskocil, N. Kriz and J. Bures, *Brain Res.* **39**, 255 (1972).
189. P. Hnik, F. Vyskocil, N. Kriz and M. Holas, *Brain Res.* **40**, 559 (1972).
190. J. Hinke, *Nature* **184**, 1257 (1959).
191. F. Vieira and G. Malnic, *Amer. J. Physiol.* **214**, 710 (1968).
192. S. Friedman, J. Jamieson,.J. Hinke and C. Friedman, *Proc. Soc. Exptl. Biol. Med.* **99**, 727 (1958).
193. S. Friedman, J. Jamieson, J. Hinke and C. Friedman, *Amer. J. Physiol.* **196**, 1049 (1959).
194. E. Moore in R. Durst, *Ion-Selective Electrodes.* Ref. 1, p. 226.
195. C. Sachs and A. Bourdeau, *J. Phys. Paris* **62**, 21 (1970).
196. D. Bernstein, M. Aliapoulios, R. Hattner, A. Wachman and B. Rose, *Endocrinology*, **85**, 589 (1969).
197. H. Quittner and W. Bridger, Division of Clinical Pathology, University of Arkansas, Medical Centre, Applied Seminar on Laboratory Diagnosis of Endocrine Disorders, held in Washington D.C., on November 7–9 (1969).
198. C. Sachs and A. Bourdeau, Annual Report of Unit No. 30, Hospital of Sick Children, Paris 1969, 12.
199. I. Radde, D. Parkinson, B. Hoffken, K. Appiah and W. Hanley, *Ann. Roy. Coll. Phys. Surg. of Canada* **3**, 34 (1970).
200. P. Venkateswarlu, L. Singer and W. Armstrong, *Anal. Biochem.* **42**, 350 (1971).
201. D. Taves, *Nature* **217**, 1050 (1968).
202. D. Taves, *Talanta* **15**, 1015 (1968).
203. L. Singer and W. Armstrong, *Arch. Oral Biol.* **14**, 1343 (1969).
204. G. Kakabadse, B. Manohin, J. Bather, E. Weller and P. Woodbridge, *Nature* **229**, 626 (1971).
205. Y. Ericsson, I. Hellstrom and Y. Hofvander, *Acta Paediat. Scand.* **61**, 459 (1972).
206. P. Grøn, K. Yao and M. Spinelli, *J. Dent. Res. Supplement to No. 5* **48**, 709 (1969).
207. A. Chernian and J. Hill, *Clin. Chem.* **17**, 652 (1971).
208. L. Kopito and H. Schwachman, *Pediatrics* **43**, 794 (1969).
209. H. Degenhart, G. Abein, B. Bevaart and J. Baks, *Clin. Chim. Acta* **38**, 217 (1972).
210. C. Sander and P. Ts'o, *J. Mol. Biol.* **55**, 21 (1971).
211. R. Coleman, *Clin. Chem.* **18**, 867 (1972).
212. J. Renfro and Y. Patel, *J. Appli. Physiol.* **37**, 756 (1974).
213. R. Kramer and H. Lagoni, *Naturwissenschaften* **56**, 36 (1969).
214. P. Muldoon and B. Liska, *J. of Dairy Sci.* **52**, 460 (1968).
215. C. Sachs, *La Presse Med.* **78**, 1547 (1970).
216. A. Perkins, M. Snyder, C. Thacher and M. Roles, *Transfusion* **11**, 204 (1971).
217. C. Sachs, A. Bourdeau and V. Presle, *Rev. Europ. Etudies Clin. et. Biol.* **16**, 831 (1971).
218. A. Bourdeau, C. Sachs, V. Presle and M. Dromini, *Pharm. Biol.* **67**, 527 (1970).
219. R. Hattner, J. Johnson, D. Bernstein, A. Wachman and J. Brackman, *Clin. Chim. Acta* **28**, 67 (1970).

220. B. Hoffken, D. Parkinson, P. Storms and I. Radde, *Clin. Orthop. and Related Res.* **78**, 30 (1971).
221. A. Bourdeau and C. Sachs, *La Pharmacieu Biologiste* **6**, 527 (1970).
222. C. Sachs, A. Bourdeau, M. Broyer and S. Balsan, *Ann. Biol. Clin.* **28**, 137 (1970).
223. C. Sachs, *Presse Medicale* **78**, 1547 (1970).
224. A. Covington and R. Robinson, *Anal. Chim. Acta* **75**, 219 (1975).
225. T. Li and J. Piechocki, *Clin. Chem.* **17**, 411 (1971).
226. J. Ruzicka and J. Tjell, *Anal. Chim. Acta* **47**, 475 (1969).
227. E. Moore, *Gastroenterology* **60**, 43 (1971).
228. O. Studer, McKnob and V. Binswanger, *Schweiz. Med. Wochschr.* **102**, 305 (1972).
229. I. Radde, D. Parkinson, B. Hoffken, K. Appiah and W. Hanley, *Pediat. Res.* **6**, 43 (1972).
230. C. Pittinger, *C.R.C. Critical Reviews in Clinical Laboratory Sciences* **1**, 351 (1970).
231. C. Fuchs, K. Paschen, P. Spieckermann and C. Westberg, *Klin. Wochenschr,* **50**, 824 (1972).
232. J. Ladensen and G. Bowers, *Clin. Chem.* **19**, 565 (1973).
233. W. Robertson and M. Peacock, *Clin. Chim. Acta* **20**, 315 (1968).
234. C. Pittinger, P. Chang and W. Faulkner, *Southern Med. J.* **64**, 1211 (1971).
235. A. Raman, *Clin. Biochem.* **4**, 141 (1971).
236. D. Killen, E. Grogan, R. Gower and H. Collins, *Surgery* **70**, 736 (1971).
237. F. Lindgarte and O. Zettervall, *Israel J. Med. Sci.* **7**, 510 (1971).
238. S. Hansen and L. Theodorsen, *Clin. Chim. Acta* **31**, 119 (1971).
239. H. Schwartz, B. McCorville and E. Christopherson, *Clin. Chim. Acta* **31**, 97 (1971).
240. A. Raman, *Biochem. Med.* **3**, 369 (1970).
241. J. Hinkle and L. Cooperman, *Brit. J. Anaesth.* **43**, 1108 (1971).
242. C. Pittinger, *Anes & Anal.* **49**, 540 (1970).
243. D. Arnold, M. Stansell and H. Malvin, *Am. J. Clin. Pathol.* **49**, 627 (1968).
244. C. Sachs, A. Bourdeau and S. Balsan, *Ann. Biol. Clin.* **27**, 487 (1969).
245. E. Moore, *J. Clin. Invest.* **49**, 318 (1970).
246. C. Sachs and A. Bourdeau, *Clin. Orthop. and Related Res.* **78**, 24 (1971).
247. C. Fuchs and K. Paschen, *Deut. Med. Wochenschr.* **97**, 23 (1972).
248. F. Lindgarde, *Clin. Chim. Acta* **40**, 477 (1972).
249. A. Raman, *Clin. Biochem.* **5**, 208 (1972).
250. S. Simard, M. Dorval, *Union Med. Can.* **102**, 551 (1973).
251. V. Subryan, M. Popvtzer, S. Parks and E. Reeve, *Clin. Chem.* **18**, 1459 (1972).
252. M. Bogunovic and Z. Stefanac, *Acta Pharm. Jugost.* **25**, 255 (1975).
253. I. Oreskes and K. Douglas, *Amer. Rheumatism Assoc. Annual Meeting* 1968, 502.
254. I. Oreskes, C. Hirsch, K. Douglas and S. Kupfer, *Clin. Chim. Acta* **21**, 303 (1968).
255. I. Radde, B. Hoffken, D. Parkinson, J. Sheepers and A. Luckham, *Clin. Chem.* **17**, 1002 (1971).
256. G. Barry, *Am. J. of Physiol.* **220**, 874 (1971).
257. L. Braddock, I. Nagelberg, E. Margallo and G. Barbero, Cystic Fibrosis Club Abstracts, April 29, 1969, 17.
258. J. Low, M. Schaaf, J. Earll, J. Piechocki and T. Li, *J. Amer. Med. Assoc.* **223**, 152 (1973).
259. M. Urist, D. Speer, K. Ibsen and B. Strates, *Calc. Tissue Res.* **2**, 253 (1968).
260. E. Moore and G. Makhlouf, *Gastroenterology* **55**, 465 (1968).
261. J. Jacobs, R. Hattner and D. Bernstein, *Clin. Chim. Acta* **31**, 467 (1971).
262. W. Robertson, *Clin. Chim. Acta* **24**, 149 (1969).
263. F. Oehme, *Glas.-Instr.-Tech.* **11**, 1162 (1968).
264. D. Armstrong and J. Kramer, Paper presented at the XII Annual Meeting of the Cystic Fibrosis Club, Atlantic City, N.J., April 28, 1971.

265. P. Muldoon and B. Liska, *J. Dairy Sci.* **54**, 117 (1971).
266. A. Hehir, C. Beck, T. Prettejohn, *Aust. J. Dairy Technol.* **26**, 110 (1971).
267. W. Krijgsman, J. Mansveld and B. Griepink, *Clin. Chim. Acta* **29**, 575 (1970).
268. C. Weber and B. Reid, *J. Nutr.* **97**, 90 (1968).
269. R. Aasenden, F. Brudevold and B. Richardson, *Arch. Oral Biol.* **13**, 203 (1968).
270. P. Grøn, H. McCann and F. Brudevold, *Arch. Oral Biol.* **13**, 203 (1968).
271. J. Birkeland, *Caries Res.* **7**, 11 (1973).
272. F. Brudevold, Y. Bakhos and P. Grøn, *Arch. Oral Biol.* **18**, 699 (1973).
273. Y. Iizuka, C. Akiba and Y. Nakayama, *Bull. of Tokyo Dental College* **11**, 155 (1970).
274. P. Hotz, H. Muhlemann and A. Schait, *Helv. Odont. Acta* **14**, 26 (1970).
275. M. Larsen, M. Kold, R.VonderFehr, *Caries Res.* **6**, 193 (1972).
276. H. McCann, *Arch. Oral Biol.* **13**, 475 (1968).
277. P. Hotz, H. Muhlemann and A. Schait, *Helv. Odont. Acta* **14**, 26 (1970).
278. R. Aasenden, E. Moreno and F. Brudevold, *Arch. Oral Biol.* **17**, 355 (1972).
279. J. Birkeland, *Caries Res.* **4**, 243 (1970).
280. F. Brudevold, H. McCann and P. Grøn, *Arch. Oral Biol.* **13**, 877 (1968).
281. P. Hotz, *Helv. Odont. Acta* **16**, 32 (1972).
282. W. Edgar, G. Jenkins and A. Tatevossian, *British Dental J.* **128**, 129 (1970).
283. L. Singer, B. Jarvey, P. Venkateswarlu and W. Armstrong, *J. Dent. Res.* **49**, 455 (1970).
284. L. Singer and W. Armstrong, *Anal. Chem.* **40**, 613 (1968).
285. A. Pochomis and F. Griffith, *Amer. Ind. Hyg. Assoc. J.* **32**, 557 (1971).
286. G. Bang, T. Kristoffersen and K. Meyer, *Acta Path. Microbiol. Scand.* 1970, Section A, **78**, 1, 49.
287. M. Sun, *Am. Ind. Hyg. Assoc. J.* **2**, 133 (1969).
288. J. Cooke and R. Hall, *Nature* **227**, 1260 (1970).
289. C. Collombel, J. Bureau and J. Cotte, *Ann. Pharm. Fr.* **29**, 541 (1971).
290. J. Tusl, *Clin. Chim. Acta* **27**, 216 (1970).
291. J. Neefus, J. Cholak and B. Saltzman, *Am. Ind. Hyg. Assoc. J.*, **31**, 96 (1970).
292. A. Cernik, J. Cooke and R. Hall, *Nature* **227**, 1260 (1970).
293. J. Tusl, *Anal. Chem.* **44**, 1693 (1972).
294. Y. Ericsson, *Acta Odont. Scandinavia* **29**, 43 (1971).
295. C. Collombel, J. Bureau and J. Cotte, *Annales Pharmaceutiques Francaises* **29**, 541 (1971).
296. L. Singer, W. Armstrong and J. Vogel, *J. Lab. & Clin. Med.* **74**, 354 (1969).
297. B. Fry and D. Taves, *J. Lab. Clin. Med.* **75**, 1020 (1970).
298. B. Panner, R. Freeman, L. Roth-Moyo and W. Markowitch, *J. Amer. Med. Assoc.* **214**, 86 (1970).
299. D. Taves, *J. Dent. Res.* **50**, 783 (1971).
300. J. Jardillier and G. Desmet, *Clin. Chim. Acta* **47**, 357 (1973).
301. P. Ke, H. Power and L. Regier, *J. Sci. Food Ag.* **21**, 108 (1970).
302. P. Ke, L. Regier and H. Power, *Anal. Chem.* **41**, 1081 (1969).
303. O. Dirks, J. Jongelius, T. Flissebaalje and I. Gedalia, *J. Dent. Res.* 1973, 588.
304. S. Manahan, *Appl. Microbiol.* **18**, 479 (1969).
305. G. Moody and J. Thomas, *J. Sci. Fd. Agric.* **27**, 43 (1976).
306. R. Khuri, W. Flanigan and D. Oken, *J. Appl. Physiol.* **21**, 1568 (1966).
307. J. Ruzica, E. Hansen and E. Zagatto, *Anal. Chim. Acta* **88**, 1 (1977).
308. H. Marsoner and K. Harnoncourt, *Aerztl. Lab.* **18**, 397 (1972).
309. W. Armstrong and C. Lee, *Science* **171**, 413 (1971).
310. R. Thomas, *J. Physiol.* **220**, 55 (1972).
311. J. Pearson and C. Gray, *J. Hosp. Pharm.* **31**, 20 (1973).

2. TYPES AND CHARACTERISTICS OF ION-SELECTIVE ELECTRODES

I. INTRODUCTORY REMARKS

Electrodes containing membranes having a specific response for a particular ion are known as ion-selective electrodes. These membranes can be conveniently classified into several categories, each category possessing certain common features.

(1) *Homogeneous solid state membrane electrodes.* The membranes of these electrodes contain only the materials responsible for the electrochemical behavior of the membranes in the form of single crystal, compact polycrystalline, or amorphous inorganic precipitate.

(2) *Heterogeneous solid state membrane electrodes.* The membranes of these electrodes consist of electrochemically active materials in the form of organic salts, inorganic salts or ion exchangers, embedded in an inert supporting material which acts as a binder. These binders are usually polymeric substances.

(3) *Glass membrane electrodes.* The membranes of these electrodes are made by melting mixtures of some elements of group 3 or 4 (e.g. Al, Si) with the oxides of some elements of group 1 or 2 (e.g. Na, Ca). These membranes permit the mobility of monovalent cations and are good sensors for hydrogen, alkali metal and alkaline earth metal ions.

(4) *Liquid membrane electrodes.* The membranes of these electrodes are composed of large organic molecules, ion exchangers, complex species, or ion-pair compounds, dissolved in water-immiscible organic solvents. Neutral sequestering substances (e.g. cyclic polyethers) may also be used as the membrane material.

(5) *Gas-sensing membrane probes.* These are electrochemical cells and not simple electrodes. The gaseous species diffuse across a gas permeable membrane, and are dissolved in an internal electrolyte, causing a change in the concentration of some ionic species. This change is detected by the electrode system.

Homogeneous and heterogeneous solid state as well as glass membranes have fixed anionic or cationic sites available at the membrane surface for ion exchange. These sites cannot move during measurement. Contrastingly, liquid membranes have mobile anionic or cationic sites. Hence liquid membranes offer a wide range of possibility for many ions, but usually have low selectivity as compared with the solid state membranes.

Solid state electrodes can measure extremely small volumes of solution [1]. Some solid state and glass electrodes and gas-sensing membrane probes can be used as highly selective sensors for many enzymatic reactions. The substrate or enzyme can be interposed between the sample solution and the sensing membrane, either by trapping in a double cellophane layer or by immobilization in a matrix of a gel or polymer, or may be applied as a coat on the electrode membrane. As a result of the reaction gaseous or ionic species are released and measured by the electrode. These types of sensors are known as "enzyme electrodes or enzyme probes".

II. HOMOGENEOUS SOLID STATE MEMBRANE ELECTRODES

The sensing membrane of these electrodes is a disk-shaped section of a single crystal or pressed pellet of fine crystals. These membranes should have the following characteristics: (i) high ionic conduction at room temperature; (ii) high mechanical stability; (iii) low solubility product; (iv) no interaction with the test solution; (v) minimum photo-electric response; (vi) good selectivity; (vii) reasonable resistivity and (viii) fast response time. These membranes may be considered as ion exchange membranes having a particularly high degree of selectivity with the advantages of long operation life, resistance to corrosion and freedom from redox interferences. When a section of a single crystal or pellet of polycrystalline material is interposed between two solutions containing one of the crystal ions, an electric potential difference is observed. One of the lattice-ions in such crystals, which is the smallest ion both in radius and charge, is involved in the conduction process. This proceeds by a lattice defect in which a mobile ion adjacent to a vacancy defect moves into the vacancy. Since the mobility of a particular ion in the crystal depends on the shape, size and charge of this ion, the conduction process is restricted to this particular ion; all other ions are unable to move in this fashion and hence cannot participate in the conduction process. For this reason, the single crystal membrane is highly selective and the interferences are usually not caused by foreign ions but are due to some chemical reactions at the crystal surface.

A list of commercially available homogeneous solid state membrane electrodes is given in Table 2.1.

Table 2.1. Types and characteristics of some commercially available homogeneous solid state membrane electrodes

Electrode	Manufacturer[a]	Model	Lower concentration limit	pH range	Interferences
Bromide	Beckman	39602	10^{-7}	0–14	I^-, S^{2-}, CN^-
	Coleman[b]	3-801	10^{-7}	0–14	S^{2-}, CN^-
	Orion	94-35	5×10^{-6}	0–14	S^{2-}
	Philips	IS 550-Br	10^{-6}	0–14	S^{2-}, Ag^+, I^-, CN^-
	HNU	ISE-30-35-00	5×10^{-7}		I^-, CN^-, S^{2-}
Cadmium	Orion	94-48	10^{-7}	1–14	Ag^+, Hg^{2+}, Cu^{2+}
Chloride	Beckman	39604	5×10^{-5}	0–14	$Br^-, I^-; S^{2-}, CN^-$
	Coleman	3-802	10^{-6}	0–14	S^{2-}, CN^-
	Orion	94-17 96-17[c] 98-17[d]	5×10^{-5}	0–13	S^{2-}
	Philips	IS550-Cl	5×10^{-5}	0–14	$S^{2-}, Ag^+, NH_3, CN^-, I^-, S_2O_3^{2-}$
	HNU	ISE-30-17-00	5×10^{-6}		$Br^-, I^-, CN^-, OH^-, S^{2-}$
Copper	Beckman	39612[e]	10^{-8}	0–14	Ag^+, Hg^{2+}
	Coleman	3-804	10^{-6}	0–14	Ag^+, Hg^{2+}
	Orion	94-29	10^{-7}	0–14	Ag^+, Hg^{2+}, S^{2-}
	HNU	ISE-30-29-00	5×10^{-7}		$Ag^+, Hg^{2+}, Cd^{2+}, Fe^{3+}, S^{2-}$
Cyanide	Orion	94-06	10^{-6}	10	S^{2-}
	Philips	IS 550-CN	10^{-6}	11	S^{2-}, Ag^+, NH_3, I^-
	HNU	ISE-30-13-00	5×10^{-7}		I^-, S^{2-}
Fluoride	Beckman	39600	10^{-6}	0–13	OH^-
	Coleman	3-803	10^{-6}	4–8	OH^-
	Orion	94-09 96-09[c]	10^{-6}	0–11	OH^-
Iodide	Beckman	39606	10^{-8}	0–14	S^{2-}
	Orion	94-53	5×10^{-8}	0–14	S^{2-}
	Philips	IS 550-I	10^{-8}	0–14	S^{2-}, Ag^+, NH_3, I^-
	HNU	ISE-30-53-00	5×10^{-8}		CN^-, S^{2-}
Lead	Orion	94-82	10^{-7}	2–14	Ag^+, Hg^{2+}, Cu^{2+}
Sodium	Orion	94-11 96-11[c] 98-11[d]	10^{-6}	3–12	Ag^+
Sulfide	Beckman	39610	10^{-9}	0–14	
	Coleman	3-805	10^{-7}	0–14	Hg^{2+}
	Orion	94-16	10^{-7}	0–14	Hg^{2+}
Thiocyanate	Orion	94-58	10^{-5}	0–14	Hg^{2+}, Cu^{2+}, S^{2-}

A. Single crystal membrane: the fluoride electrode

The unique properties of the rare earth fluorides (e.g. lanthanum, neodymium and praseodynium fluorides) to form a membrane exclusively permeable to fluoride ions gave rise to the first single crystal membrane specific to fluoride. This is one of the few truly specific ion-selective electrodes [2]. The lanthanum fluoride crystal has an electrical conductance [2a] of about $10^7 \, \Omega^{-1} \, cm^{-1}$ at 25°C. The mobility of the fluoride ion in the crystal lattice has been shown by the narrowing of the ^{19}F n.m.r spectra [2a] as the temperature increases in the region of 300 K.

The conductance of pure rare earth fluorides can be increased by doping the crystal with divalent ions such as Eu^{2+} or Sr^{2+}. Lingane [3] has reported that the quantity of Eu^{2+} used to dope the lanthanum fluoride membrane to create fluoride holes, and whether or not this really is necessary, remains to be clarified. Since EuF_3 ($K_s \, 2 \cdot 2 \times 10^{-17}$) is more soluble than LaF_3 ($K_s \sim 10^{-29}$), the addition of large quantities of europium should lower the limit of fluoride to which the membrane would respond. Presumably, therefore, only a very small quantity of europium is used to improve the crystal conduction, but this is not essential for the functioning of the electrode. Single crystal membranes which are selective for ions other than the fluoride are also known. For instance, a synthetic chalcocite (Cu_2S) single crystal in the form of a plate (1 mm thick and 5 mm diameter) mounted at the end of a glass tube by means of silicone rubber has been used as a membrane for copper [4].

The construction of the fluoride electrode is similar in principle to that of a conventional glass pH electrode, except that the membrane material is a disk-shaped section of a single-crystal of lanthanum fluoride. The membrane (1 cm in diameter and 1–2 mm thick) is sealed with epoxy cement to the end of a hollow rigid poly (vinyl chloride), polypropylene or Teflon tube which is electrically insulated and chemically inert [2] (Fig. 2.1). The

---

[a] Philips: Eindhoven, Netherlands; (Pye Unicam) York Street, Cambridge, U.K.
Coleman: 42 Madison Street, Maywood, Illinois 60153, USA. Perkin-Elmer Ltd., Beaconsfield, Bucks, U.K.
Beckman: 2500, Harbor Boulevard, Fullerton, California 92634 USA. Glenrothes, Fyfe, Scotland, PO Box 1, U.K.
Corning: Medfield, Mass. 02052 USA. 3 Cork Street, London W1, U.K.
Orion: 11 Blackstone Street, Cambridge, Mass. 02139, USA. E.I.L., Richmond, Surrey, U.K.
HNU: HNU Systems, 30 Ossipee Road, Newton Upper Falls, Mass. 02164, USA. Via Boccaccio, 2, Milano 20123, Italy.
[b] No internal filling solutions are used in these electrodes.
[c] Combination electrodes.
[d] Flow-through electrodes.
[e] Like Coleman electrodes.

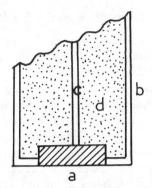

FIG. 2.1. Solid state macro-electrode for sample volumes above 1 ml. a, electrode membrane; b, electrode body; c, internal reference electrode; d, internal filling solution.

only critical step is to seal the membrane tightly. The early commercial models of this electrode were made by holding the crystal membrane in place by pressure wedge sealing. Methods for casting the crystal into a strong viscous resin able to withstand corrosion and resist "memory effects" have been described [5]. The combined casting is then cemented into the shaft on the electrode. Flushing the external membrane surface with the polymer body prevents occlusion [6]. The electrode tube is filled with a solution containing both fluoride and chloride ions ($0 \cdot 1$ M of sodium fluoride and $0 \cdot 1$ M potassium chloride) and an internal Ag/AgCl reference electrode is immersed in the filling solution. The internal electrical connections may be directly bound to the inner surface of the crystal using a metal-impregnated epoxy resin [7].

Many types of these membrane electrodes are commercially available. These are Orion (94-09), Coleman (3-803), Corning (476042) and Beckman (39600). Most of these withstand mechanical and thermal shocks without cracking or breaking and can be used at temperatures between −5 and 100°C, but it should not be subjected to a sudden change in temperature of more than 50°C, in order to avoid damaging the seal between the membrane and the body of the electrode. Such electrodes are inert to 10% sodium hydroxide, 25% nitric acid, 10% hydrofluoric acid, methanol, benzene, acetonitrile, ethyl alcohol, glacial acetic acid, acetone and dioxane, but must not be used in dimethylformamide, chloroform or other strong polar solvents.

These electrodes can be used with sample volumes of 5 to 50 μl by inverting the electrode and using the membrane itself as sample container [8]. Disconnection between the inner surface of the membrane and the electrode solution, which may arise by entrapment of air bubbles when the electrode is inverted, is alleviated by using agar gel inside the electrode. The electrode is opened by placing it in a boiling water-bath, screwing the top and removing the inner ring seal. A gel composed of 4% agar, 0·1 M NaF and 0·1 M KCl is heated and transferred by means of syringe to the inner electrode compartment. It is not necessary to fill the compartment completely. The electrode is then reassembled and allowed to cool slowly in the normal vertical position. On operation, the electrode is inverted and a Tygon tubing (plasticized PVC-6 mm i.d.) sleeve which forms a very tight seal with the raised position of the membrane is used. A silicone lubricant is applied between the crystal membrane and the sleeve (Fig. 2.2).

Although this modification accommodates the use of small sample volumes, they are not adaptable for *in vivo* and certain *in situ* measurements. Durst [9], however, has described a micro-electrode suitable for these purposes. A small cone-shaped piece of lanthanum fluoride crystal [10] is heat sealed into a polyethylene tube (2 mm o.d.). The tip of the crystal is painted with polystyrene coil dope, which after drying is scraped away to expose a small, well defined conical portion of approximately 1·5 μl volume of the crystal (Fig. 2.3). This small area of exposed membrane can be reduced by shortening the length of the embedded lanthanum crystal. A silver–silver chloride internal reference electrode and sodium fluoride–potassium chloride internal solution are used as in the macro version of the electrode. A combination electrode in which both the fluoride and reference sensing elements are built within the same electrode body may also be used (Fig. 2.4).

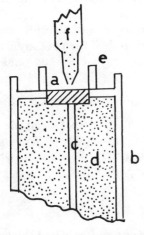

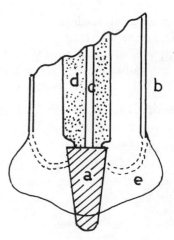

FIG. 2.2. Inverted macro-electrode for sample volumes above 50 μl. a, electrode membrane; b, electrode body; c, internal reference electrode; d, internal filling solution with agar gel; e, Tygon sleeve; f, reference electrode.

FIG. 2.3. Micro-electrode for sample volumes above 2 μl. a, electrode membrane; b, electrode body; c, internal reference electrode; d, internal filling solution; e, polystyrene coil dope. (Courtesy of the American Chemical Society.)

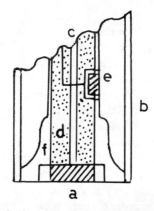

FIG. 2.4. Combination ion-selective electrode for sample volumes above 10 μl. a, electrode membrane; b, electrode body; c, internal reference electrode; d, internal filling solution; e, combined reference electrode; f, conical leak for liquid–liquid junction.

A lanthanum fluoride membrane electrode for use in molten fluoride salts [11] has been described. The lanthanum fluoride crystal is supported in a nickel tube to prevent undue dissolution and the contact with the molten test samples of fluoride is made through a porous nickel frit. Electrical continuity is made by a mixed fluoride melt contained in a small hole drilled in the top of the crystal, into which a nickel contact is suspended. The electrode response is stable, barely soluble in the melt and yields accurate e.m.f. readings.

The principal application of the electrode is to measure fluoride activity or concentration in aqueous solutions. The electrode is used in conjunction with a saturated calomel electrode, as a reference electrode, and the potential is developed from the cell:

$$Hg;\ Hg_2Cl_2(s)|KCl(sat.)\|F^-\|LaF_3(s)|NaF(0\cdot1\ \text{M});\ NaCl(0\cdot1\ \text{M})|AgCl(s);\ Ag$$

The half-cell to the left is a commercial saturated calomel electrode and the half cell to the right is contained within the plastic body of the fluoride electrode. Since the membrane is only permeable to the fluoride ion, whose activity in the internal solution is constant, the potential of the cell is given by the equation:

$$E = E_{\text{AgCl;Ag}} - E_{\text{SCE}} - RT/\text{F} \ln [\text{F}^-]\nu_{\text{F}} + E_{\text{a}} + E_{\text{j}} \qquad (1)$$

where E_{a} is the asymmetric potential of the membrane and E_{j} is the liquid junction potential between the saturated potassium chloride solution and the test solution. Both of these terms are usually combined with the potential of the internal Ag/AgCl electrode and the calomel reference electrode and are found to be relatively constant in time and relatively independent of the composition of the test solution.

$$E = E_0 - RT/F \ln [F^-] \nu_F \qquad (2)$$

or

$$E = E_0 + 0.05915 \, pF^- \qquad (3)$$

where pF is the negative logarithm of the fluoride ion activity in the test solution and the constant E_0 is the sum of the potentials of the Ag/AgCl electrode, the saturated calomel reference electrode, the liquid junction potential between the test solution, and the reference electrode and the potential across the membrane when the fluoride ion activity in the test solution is unity. In a constant ionic medium the actual value of the activity coefficient need not be known since it is constant throughout the various experiments and can be included in the E_0 value when the calibrations are made. Potassium fluoride is preferred to sodium fluoride [12] for electrode calibration because of its greater solubility and freedom from ion association.

1. *Effect of Interfering Ions*

The specificity of the fluoride electrode in the presence of chloride and nitrate ions has been studied by many workers [2, 8, 13–14a]. It has been found that the addition of 0·1 M sodium chloride or sodium nitrate to a solution containing 10^{-4} M fluoride ion results in a 7 mV decrease in fluoride potential. This is expected from the change in the ionic strength. In the presence of 1 M chloride or nitrate the potential of the fluoride solution $(10^{-4}$ M) is depressed by about 20 mV. The response of a fluoride electrode down to 2×10^{-7} M fluoride in pure solution as well as in 1 M sodium chloride background [15] is Nernstian if the readings are corrected for the difference in fluoride ion activity coefficients and liquid junction potentials. In general, chloride ions do not interfere up to the ratio of $5·6 \times 10^6 : 1$. In the presence of sulfate ions (e.g. 1 M solution), the potential of the electrode in 10^{-4} M fluoride only shows activity suppression equivalent to 13 mV rather than the predicted 18 mV for monoelectrolytes [2]. This is certainly within the possible error of estimating the activity coefficient of fluoride in the presence of a divalent ion. The phosphate ion shows no interfering effect in the acidic range [16–19]. In a citrate buffer [20], or acetate buffer solution [21], the fluoride potential is not affected. For the potentiometric determination of fluoride in the presence of phosphate, using lanthanum nitrate as a titrant at pH 5–7, the interference due to phosphate (up to ten-fold in excess) can be tolerated by adsorption on zinc oxide [22]. The latter is added in excess and need not be removed before titration. Fluoride potential in the presence of a 65-fold excess of silicates shows a negative shift by only 0·3 mV. The potential becomes 5 mV more positive when the pH changes from 7 to 4. This is probably due to the formation of a slightly charged film of silicate-based material on the surface of the lanthanum fluoride membrane. A fluoride ion in the form SiF_6^{2-} gives only slightly less response than the equivalent amount of fluoride ion [23].

In the presence of cations that form complexes with the fluoride ion [24], the lanthanum fluoride electrode responds theoretically to the free fluoride ion only below 10^{-9} M. In the presence of some cations that undergo reactions with the fluoride ion, such as aluminium and ferric ions, serious interferences due to the formation of metal fluoride complexes are observed. However, with these two particular cations, raising the pH to about 7 would obviate this difficulty by precipitating their hydroxides and therefore reducing their fraction present in the ionic form. The effect of aluminium can also be removed by using a sodium acetate–potassium nitrate mixture [25]. On the other hand, the electrode responds to the calcium ion; after exposure to solutions containing calcium, it requires prolonged washing to restore its original conditions [26].

The only significant interfering anion is the hydroxide ion. This interference may be understood in view of the similarity of the hydroxide and fluoride ions both in charge and ionic radii. In pure solutions, significant interference occurs when the concentration of the hydroxide ion is equal to that of the fluoride. A ten-fold excess of hydroxide will double the apparent fluoride content [2]. In most applications, adjustment of the pH can easily be made before measurement.

Carboxylate ions (e.g. formate, acetate, oxalate, lactate, propionate and butyrate) do not seriously affect the normal operation of the fluoride electrode [16, 27, 29]. The infrared studies showed no evidence of the formation of a layer of insoluble lanthanum fluoride-carboxylate or lanthanum fluoride precipitate [29]. When the carboxylate of a strong ligand tendency (e.g. acetate, malonate, citrate or acetylacetonate) is present, the fluoride ion level produced by the reaction with the membrane is small compared with the fluoride in the test solution and can be ignored. However, high concentration of these complexing agents (e.g. citrate ion) dissolves the membrane [28].

$$LaF_3(s) + Cit^{-3} \rightleftharpoons LaCit + 3F^- \tag{4}$$

The response of the electrode in non-ionic organic compounds was also tested. No evidence was found for fluoride binding by albumin, globulin, lysozome, dextran, and insoluble polysaccharides [30]. Glucose and carbamide cause only a slight shift in the calibration graph. It was also found [26] that the fluoride electrode gives a response to 1-fluoro-2,3-dinitrobenzene (0·16 to 8 mM) in buffered and unbuffered aqueous solutions although no free fluoride ions are present.

While the fluoride electrode is probably the most intensively studied, other electrodes also have received attention. For instance, Havas and Kecskes [30a] have evaluated capillary micro- and macro-chloride sensitive electrodes.

B. Compact polycrystalline membranes

1. *Halide Selective Membranes*

Membranes prepared from a single crystal or pressed disk of silver halides possess high electrical resistance and large photoelectric potential; hence it is necessary to use them under constant illumination conditions. Furthermore, some silver halides (e.g. silver iodide) when pressed undergo several solid phase transformations causing the membrane to fracture. These drawbacks are overcome by annealing the silver chloride single crystal at 320°C for 40 h at reduced pressure and allowing the membrane to cool at a controlled rate [31]. Another technique utilizes the spontaneous reaction of silver chloride with aqueous iodide to form silver iodide [32, 33] membrane.

When silver halides are mixed with silver sulfide, they produce ionic conductors in which silver is the mobile species. These are used as active materials in the fabrication of many commercially available electrodes. Halide membranes prepared from a 1:1 mol ratio of silver sulfide and silver halide give the best results, while other ratios down to 1:9 produce fairly good sensors [34].

The halide membrane materials are prepared [35] by co-precipitation of a 50% mol ratio of silver sulfide and silver halides. The stoichiometric amount of silver nitrate is added to the equimolar mixture of sulfide and halide. The precipitate is washed with hot water and acetone and then dried in air at 100°C. Two grams of the precipitate are pressed using a potassium bromide die in a vacuum at room temperature for one hour under a 20,000 pound load (100,000 p.s.i.). The membrane produced (~2 mm thick) is mounted on a tube filled with 0·1 M of the halide solution and a silver/silver chloride electrode is immersed in the internal reference solution. The chloride membrane produced by this technique behaves as a true mixture of silver chloride and silver sulfide, whereas the bromide and iodide membranes are probably Ag_3SBr and Ag_3SI compounds, respectively. The latter two membranes show no evidence of cracking since the membrane compounds do not exhibit a phase transition or the rate of such transition is too slow to be observed.

The potential developed by the halide solutions and sensed by the silver halide/silver sulfide membrane electrodes is based on the change in silver ion activity which depends, in turn, on the halide ion activity [36].

$$E = E_a + 2·303 \, RT/F \log a_{Ag^+} \tag{5}$$

The silver ion activity can be calculated from the solubility product (K_{sp}) of silver halide.

$$a_{Ag^+} = \frac{K_{sp}}{a_{x^-}} \tag{6}$$

Then

$$E = E_0 + 2 \cdot 303 \, RT/F \log K_{sp} - S \log a_{x^-} \tag{7}$$

Since K_{sp} is constant at any given temperature, the above equation becomes:

$$E_b = E_0 + 2 \cdot 303 \, RT/F \log K_{sp} \tag{8}$$

$$E = E_b - 2 \cdot 303 \, RT/F \log a_{x^-} \tag{9}$$

The halide electrodes can be used in solutions containing oxidizing agents such as cerium (IV) and manganate ions [36, 37]. Thus, it is possible to determine the chloride ion in the presence of high amounts of bromide ions after selective oxidation of the bromide with either potassium permanganate or ceric sulfate [38]. The halide electrodes are also applicable in measuring halide ion activity in non-aqueous media [39].

However, the halide electrodes cannot be used in the presence of strong reducing agents or species which are capable of forming more insoluble silver salt than the silver halides. For example, the thiocyanate ion interferes with the bromide membrane electrode when the CNS^- : Br^- ratio in the test solution exceeds the value given by the solubility products of AgCNS and AgBr. Under these conditions, the bromide membrane is converted to the thiocyanate and the electrode becomes selective to thiocyanate rather than the bromide ion [40].

$$SCN^- + AgBr(membrane) \rightleftharpoons AgCNS + Br^- \tag{10}$$

Similarly, copper ions at pH 2·4 severely limit the Nernstian response of the chloride electrode probably due to the formation of an insoluble layer of cuprous sulfide on the detector membrane [41].

Solid state halide ion-selective electrodes have been prepared by covering support materials such as Ag_2S, Ag_3SBr, Ag_3SI and $Ag_{19}I_{15}P_2O_7$ with a thin layer of silver halide by exposure to halogen gas or vapour, or by electrochemical deposition [42]. Membranes for chloride [43] and bromide [44] may also be prepared by compressing a $1:2$ w/w mixture of mercuric sulfide and mercurous halide under $8000 \, kg/cm^2$ at $25–200°C$. It has been reported that such membranes respond to the halide ion in the μM to M range and give better over-all performance than the electrodes based on silver halide precipitates. Vlasov et al. [44a] have fabricated a chloride and bromide responsive ion-selective field effect potentiometric sensor.

2. Cyanide and Thiocyanate Selective Membranes

The iodide electrode (silver sulfide–silver iodide membrane) can be used as a cyanide sensor [45, 46]. The response of the electrode is due to a replacement reaction whereby the iodide ion is liberated at the membrane surface [47]:

$$\text{AgI (membrane)} + 2\,\text{CN}^- \rightleftharpoons \text{Ag(CN)}_2^- + \text{I}^- \tag{11}$$

The ions released by this reaction fix the activity of silver ion at the membrane surface through the solubility product relationship:

$$a_{\text{Ag}^+} = \frac{K_{\text{sp}}}{\nu a_{I^-}} \tag{12}$$

where ν depends on the stirring rate. Since the amount of iodide liberated at the electrode membrane is proportional to the CN^- content of the test solution, it can be shown that

$$E = E_0 + 2 \cdot 3RT/F \log 2\, K_{\text{sp}}/\nu a_{\text{CN}^-} \tag{13}$$

$$E = [E_a + 2 \cdot 3\, RT/F \log 2\, K_{\text{sp}}/\nu] - 2 \cdot 3\, RT/F \log a_{\text{CN}^-} \tag{14}$$

$$E = E_b - 2 \cdot 3\, RT/F \log a_{\text{CN}^-} \tag{14a}$$

The silver sulfide of the membrane forms a permeable diffusion barrier which under equilibrium conditions enable a stable reading to be obtained by stabilizing the iodide concentration at the membrane surface. This takes place when the flux of the iodide out of the diffusion barrier is equal to one half the rate of diffusion of cyanide into the membrane. The rate of production of iodide is one half the rate of removal of the total concentration

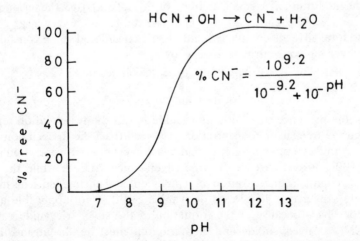

FIG. 2.5. Effect of pH on the ionization of hydrocyanic acid. (Courtesy of Orion Research Inc.)

of the cyanide ion per unit time [48]. Assuming no ion adsorption, then a stable electrode response is expected when the rate of iodide production is equal to its rate of diffusion into the outer solution [49].

The influence of pH on the electrode response can be predicted from a logarithmic diagram of the HCN/CN$^-$ system [50] (Fig. 2.5). The optimum pH range is from 12–13. Below pH 12, association of CN$^-$ with H_3O^+ occurs. However, the electrode can be used at pH values as low as 10 provided that the pH is carefully controlled [51]. The effect of pH on the electrode response is given by the relation [52]:

$$E = E_0 - 2 \cdot 3\, RT/F[pK_{HCN} - \log C_T - \log (C_{H_3O^+} + K_{CN^-})] \quad (15)$$

where C_T is equal to the total amount of HCN and CN$^-$ ions and K_{HCN} is the dissociation constant of HCN.

Whereas no interferences due to chloride and bromide ions occur up to the ratio 10^6 and 10^3, respectively, iodide ions interfere seriously [50]. Ions forming extremely stable complexes with silver (e.g. NH_3, $S_2O_3^{2-}$, S^{2-}) interfere. However, it is possible to remove the effect of the sulfide ion by prior precipitation with cadmium (II), lead (II), bismuth (III) or o-hydroxy-mercuribenzoic acid [53–55]. The excess cadmium (II), lead (II) and bismuth (III) ions do not influence the electrode response, because their complexes with cyanide are weaker than silver cyanide. It may be mentioned that under the alkaline conditions normally used for the cyanide measurments, cadmium (II) and bismuth (III) are present as hydroxides and lead as plumbite.

The most serious drawback of the use of the cyanide electrode for cyanide measurement is the slow dissolution of the membrane. This can be overcome by using a silver/sulfide electrode and an indicator technique which depends on the addition of a small amount of $K[Ag(CN)_2]$ to the cyanide test solution [56, 57]. The silver complex is so stable that a very large fraction of the total silver is complexed and can be considered to be constant; the free silver ion concentration is very low.

$$Ag(CN)_2^- \rightleftharpoons Ag^+ + 2\, CN^- \quad (16)$$

$$[Ag^+] = Constant/[CN^-]^2 \quad (17)$$

Thus, the presence of additional cyanide ion in the test solution tends to reduce the free silver concentration produced from the dissociation of the silver cyanide complex. A ten-fold increase in cyanide concentration results in a 100-fold decrease in the free silver concentration, while the silver cyanide complex concentration remains essentially unchanged. This method is quite sensitive to low levels of cyanide but it is not applicable at high cyanide levels because of the formation of the silver tricyanide complex $Ag(CN)_3^{2-}$. Silver, sulfide and mercuric ions must be absent and the pH should be adjusted to about 13.

The thiocyanate sensor is a silver thiocyanate–silver sulfide membrane which is an ionic conductor for silver. The potential developed by this membrane is due to change in the silver ion activity [58] (equation 5). Small solubility of silver thiocyanate membrane gives silver ion activity that can be calculated from the solubility product.

$$a_{Ag^+} = K_{sp}/a_{SCN^-} \tag{18}$$

$$E = E_0 + 2 \cdot 3 \, RT/F \log K_{sp} - 2 \cdot 3 \, RT/F \log a_{SCN^-} \tag{19}$$

$$E = E_b - 2 \cdot 3 \, RT/F \log a_{SCN^-} \tag{20}$$

The measurement is conducted at pH 2–10 for 10^{-3} M thiocyanate solution and for more dilute solutions a nearly neutral pH is recommended. The electrode should not be used in reducing solutions, and in solutions containing species that form silver complexes or insoluble silver salts, or in the presence of cations which form complexes with CNS^- (e.g. silver, cuprous, cupric, mercuric and ferric ions).

3. Metal Ion-Selective Membranes

Membranes selective to metallic ions are prepared by incorporating the metal sulfide with silver sulfide. The solubility of the metal sulfide must be low enough to resist membrane dissolution but should be higher than that of silver sulfide in order to maintain a layer of silver sulfide on the membrane surface.

$$Ag_2S(s) + M^{2+} = 2\, Ag^+ + MS_{(S)} \tag{21}$$

Fortunately, the last factor is satisfied by the fact that all the metal sulfides are more soluble than the Ag_2S [59].

Since the membrane contains sufficient silver sulfide to provide silver ions, it will function as a silver ion detector. Assuming that the test solution contains no silver ion, the level of silver ion in the test solution is due to membrane solubility.

$$a_{Ag^+} \cdot a_{S^{2-}} = K_{sp} \, Ag_2S \tag{22}$$

$$a_{M^{2+}} \cdot a_{S^{2-}} = K_{sp} \, MS \tag{23}$$

$$a_{Ag^+} = \sqrt{(a_{M^{2+}} \cdot K_{sp} \, Ag_2S/K_{sp} \, MS)} \tag{24}$$

Consequently, the potential developed by these electrodes is given by the equation:

$$E = E_0 + 2 \cdot 303 \, RT/nF \log a_{M^{n+}} \tag{25}$$

Membranes selective to silver, copper, cadmium and lead are prepared by co-precipitation and pressing the metal sulfide and silver sulfide under suitable conditions. Precipitation of each sulfide alone, followed by mechanical mixing in the desired ratio and pressing, may be suitable in some cases but the most general and reliable technique as recommended in the Orion patents involves co-precipitation.

The copper selective membrane may be prepared by co-precipitation of a 50% mol ratio of copper sulfide and silver sulfide and pressing the precipitate at room temperature under approximately 100,000 p.s.i. for about 30 min. The exact composition of the co-precipitate is not a critical factor and variations in the pressure and duration of pressure do not significantly affect the electrode performance [35]. This electrode responds to copper concentration over the range of 10^{-6}–10^{-9} M. The response and rate of approach to the steady state potential depends on the method of cleaning [60]. The electrode can be used for the titration of many metal ions (e.g. Zn, Fe, Mn, Cd, Ni and other transition metals) either by addition of EDTA and titration of the excess with copper solution [61] or by direct titration with EDTA using the copper ion as indicator [62]. The use of the electrode for complexometric titration of a number of metal ions in the absence of copper at pH 12 in ammoniacal buffer has been also suggested [63]. The response of the electrode is probably due to complexation equilibria of the cupric ions at the membrane surface and the titrant. On the other hand, the electrode shows a well defined response to species that form complexes with copper ion. Thus, citrate, tartrate and acetate show rectilinear relationships between their concentrations, in the range of 10 mM to μM, and the electrode potential [64].

Chloride ion significantly interferes when its concentration is equal to, or greater than, that of the copper and the membrane is tarnished, but may be restored by polishing with "crocus" emery cloth [65, 66]. The response of the electrode to copper ions in chloride background is linear with the logarithm of copper activity but not Nernstian [67]. Thus, it is recommended to use a double junction reference electrode with $1-2$ M KNO_3 filling solution in conjunction with the copper electrode. Iron (III) and Hg (II) are the main interfering cations [68] and should be absent. Iron (III) is easily removed by adjusting the pH of the solution to be >4 and <6. However, it has been reported that the copper ion-selective electrode responds also to iron (III) ions [69] in the range of 10 μM to 10 mM solutions at pH 2 in $0\cdot1$ M $NaClO_4$. The working life of the electrode as an iron (III) sensor is shorter than when used to detect Cu^{2+}. The measurement should be conducted in the dark and salicylaldoxime can be used to reduce the interferences caused by copper. Oxidizing solutions produce pits at surface dislocations resulting in mixed electrode potential, besides decreasing the slope stability and speed of response [70]. The electrode is suitable for use in non-aqueous solvents [71, 72].

A copper electrode (AgS-CuS membrane) with a 25 μl cell has been developed for the micro-determination of copper, with the use of a 5-μl burette (calibrated to 1 nl) constructed from a syringe, by standard addition and subtraction techniques [73]. A flow-through electrode system for copper measurement in the millimolar to sub-micro molar concentration range has been described [74]. A membrane selective to copper has also been prepared from a cuprous sulfide sintered disk [75, 76]. The characteristics of such an electrode depend on the ratio of the two cuprous sulfide phases present which can be controlled during the preparation of the disk. Precipitation of copper sulfide on a silver sulfide surface has been reported [77].

Lead-selective membrane is prepared in a similar way to the copper membrane but with greater difficulty. Membranes containing more than 20% lead sulfide show poor pressing characteristics and tend to crack, while membranes containing less than 20% lead show no response to lead ion. Suitable membranes are prepared by co-precipitation of a $1:1$ or $1:2$ mole ratio of lead sulfide and silver sulfide followed by thorough washing with acid, carbon disulfide and acetone before drying at 80–100°C and pressing [35, 78]. Simultaneous heating at 150–250°C and pressing at 115,000 p.s.i. for 4–8 h are recommended. Cadmium sensing membrane is prepared in a similar manner using a cadmium sulfide and silver sulfide mixture in the ratios of $1:1$, $1:2$ [35] or $1:10$ [79].

Lead and cadmium electrodes do not respond to anions and most cations. Copper, mercury and silver ions, however, poison the electrode sensing element and must be absent from the sample. Ferric or lead ions in amounts exceeding that of cadmium in the test solution interfere [80, 81]. Cadmium and lead electrodes are suitable for applications in non-aqueous media [82] (4:1 and 1:1 water–methanol or water–dimethyl sulfoxide). Evaluation of some commercial cadmium and lead electrodes in some buffered and non-buffered solutions has been reported [83, 84].

The solid state iodide electrode (AgI/Ag_2S membrane) can be used as a sensor for some ions other than the iodide ion. It responds to mercuric ion down to 10^{-8} M (2 p.p.b. of Hg^{2+}) and the electrode can also be used as a sensitive electrode for mercuric ions [85, 86]. Although the mechanism of the electrode functioning is not yet fully understood, it is possible that mercuric ions react with silver iodide on the electrode membrane surface to release silver which is sensed by the electrode.

$$Hg^{2+} + AgI \text{ (membrane)} \rightarrow HgI^+ + Ag^+ \qquad (26)$$

The measurement of mercuric ions is conducted in background solutions of 0·1 M sodium perchlorate, 0·1 M sodium nitrate, $1-2$ M nitric acid and $0·5-1$ M sulfuric acid solutions [85, 87]. The best measurements can be performed at pH 4–5. High pH causes the mercury to precipitate as hydroxide. Silver and copper ions are the only two interfering cations. The iodide electrode should not be left in mercury-containing solutions for a long time. If it is exposed to a concentrated mercury solution, it should be dipped in a solution of about 10^{-3} M sodium iodide for 10 s and rinsed with water before use for dilute mercury solutions. The electrode responds to some extent to Hg_2^{2+} ion and organic mercury compounds of the type HgR^+ but does not detect compounds of the type HgR_2.

The electrode was also found to respond to gold [88] in the range of 10^{-4} M to 10^{-8} M at pH 1·2–2·3. The electrode response is not Nernstian. Above 10^{-5} M Au^{3+}, the membrane surface becomes poisoned and requires restoration by soaking in 0·1 M KI solution. It is possible to presume exchange process on the surface of the membrane of the electrode to explain its response to Au^{3+}:

$$Au^{3+} + 3\,AgI \rightarrow AuI_3 + 3\,Ag^+ \tag{27}$$

Solid state electrodes selective for copper, lead, cadmium and mercury have been prepared by covering Ag_2S, Ag_3SBr or Ag_3SI with a thin metal sulfide layer and the latter is treated by heating under a partial pressure of sulfur [89].

4. Sulfide, Sulfite and Sulfate Selective Membranes

The sulfide membrane is composed of a silver sulfide sensor and it responds to either silver [90] ion or sulfide [91] ion in aqueous solutions. For sulfide measurement [59], the amount of silver generated by the membrane (equation 5) depends on the sulfide ion activity of the test sample. Thus the potential developed depends on the sulfide concentration:

$$E = E_0 - 2·303\,RT/2F\,\log a_{S^{2-}} \tag{28}$$

E_0 includes a factor involving the solubility product of silver sulfide. Since the membrane surface is dense and non-porous, it responds quickly to changes in sample sulfide levels as low as 10^{-18} M. Chloride, bromide, iodide, thiocyanate, chromate, nitrate and sulfate ions do not affect the response of the electrode. Mercuric ion is the only cationic interference. It should be noted that the sulfide electrode can be used for the measurement of the cyanide ions [92], and it has some advantages over the cyanide electrode: (a) it has a lower detection limit; (b) the greater slope of its Nernstian graph means more precise measurement; (c) it has a longer life; (d) it is less subject to interference by iodide ion; (e) it is useful for continuous measurement.

The sulfite ion-selective membrane is composed of a polycrystalline mixture of mercuric sulfite and mercurous chloride [93]. The response mechanism of this membrane probably involves the reaction.

$$Hg_2Cl_2 + 2\,SO_3^{2-} \rightarrow Hg + Hg(SO_3)_2^{2-} + 2Cl^- \tag{29}$$

The limit of sulfite detection is below parts per million.

The sulfate membrane [94] is prepared by compressing a mixture of silver sulfide, lead sulfide, lead sulfate and copper sulfide in the ratio of $32:31:32:5$ mol %, respectively. A pressure of 102,000 p.s.i. for 18 h at a temperature of 300°C is applied. This elevated temperature appears to be the critical variable and may be related to the fact that silver sulfide undergoes a phase transition at 178°C. The response time is improved by the presence of copper sulfide by virtue of its semiconductor properties. The sulfate electrode shows a Nernstian slope (29 mV per decade) over a wide range of sulfate activity at pH 3 to 10. It displays a good selectivity for sulfate ion over a variety of common univalent anions and is applicable to the activity measurement of sulfur-35 in aqueous sulfate solutions [95].

C. Chalcogenide membranes

It should be recognized that many metal sulfides are not suitable for membrane preparation. For example, zinc sulfide in hygroscopic, manganese sulfide is very soluble in neutral and acidic solutions, cobalt sulfide shows no response, and nickel sulfide shows poor response. This stimulated many workers to prepare chalcogenide membranes. Metal selenides and tellurides when incorporated with silver sulfide matrix function in a manner similar to that of the corresponding metal sulfide electrode, with the advantage that they are less soluble than the corresponding metal sulfides and can resist abrasion. However, electrodes based on a sulfide matrix are superior to those based on selenide or telluride and have low redox sensitivity [96].

Ion-selective chalcogenide electrodes responsive to silver [97, 98], lead [97, 99, 100], chromium [97, 101], nickel [97], cobalt [97, 102], cadmium [97, 103], zinc [97, 104], copper [97, 105–107], and manganese [108] have been prepared by using the corresponding metal chalcogenides with silver sulfide. Metal selenides or tellurides are made by direct reaction of the metal with selenium or tellurium in a high vacuum sealed quartz tube and the matrix material is prepared by reaction of silver with sulfur in a stream of hydrogen sulfide gas. The mixture of metal selenide or telluride and silver sulfide is compressed at a pressure of 10–20 ton/cm^2 to form a tablet with 2–3 mm thickness, and then sintered at 100–600°C for 3 h in a stream of an inert gas. The disk is fastened directly to a lead wire after its surface has been polished with diamond paste. The membrane is then mounted with epoxy resin adhesive on a stem in an epoxy tube filled with the electrolyte containing the metal ion of interest.

Chalcogenide glass membranes have also been prepared. A membrane of the composition $Ge_{28}Sb_{12}Se_{60}$ doped in silver chloride is selective for silver ion [109]. Sensors responding to copper and ferric ions are similarly fabricated by doping the chalcogenide glass with the corresponding metals as active material [110]. Arsenic trisulfide chalcogenide glass containing

$Cu_6As_4S_9$ as an active constituent responds to copper ion [111]. Chalcogenide glass membranes with the composition $Fe_nSe_{60}Ge_{28}Sb_{12}$ are responsive to nitrogen dioxide (NO_2) in the gaseous phase [112]. These sensors appear to act by a redox potential mechanism rather than by an ion-exchange process.

D. Ceramic membranes

These membranes are prepared by sintering and hot pressing of powdered metal salts. The membranes are sealed into the bottom of a plastic tube, attached to a wire and used without an internal electrode or reference solutions.

Lead [113], copper [113], cadmium [114], and silver [115] selective membranes are prepared by heating a mixture of the corresponding metal sulfides with silver sulfide at temperatures of 200–500°C for several hours under a pressure of 3–7 tons/cm^2.

Non-sulfide ceramic membranes are also available. Ceramic membranes selective to fluoride ions are prepared by sintering lanthanum fluoride, europium fluoride and calcium fluoride at temperatures above 1200°C for 3–15 h under a flow of hydrogen fluoride gas [116]. A membrane prepared from powdered lanthanum fluoride drifted with 0·1–0·15 mol % of europium fluoride heated *in vacuo* for 4 h and compressed under a pressure of 120–130 kg per cm^2 for one hour at 1000°C has also been described [117].

E. Organic membranes

Membranes selective for some cations and anions are prepared by compressing a suitable organic salt into disks. Silver and copper selective membranes are prepared using the silver and copper salts, respectively, of 7,7,8,8-tetracyanoquinodimethane [118, 119]. Copper and lead salts of tetraphenylarsonium-2,6-bi(dicyanomethylene) naphthalene and 9-dicyanomethylene-2,4,7-trinitro and 2,4,5-tetranitrofluorene are electrochemical sensors selective for copper and lead ions [120]. The operational range of these membranes is wider and the selectivity is better than those prepared from 7,7,8,8-tetracyanoquinodimethane salts.

A perchlorate selective membrane is prepared from the perchlorate salt of the azoviolene form of N-ethylbenzothiazole-2,2′azine [121]. The perchlorate salts of several aromatic diamines (e.g. benzidine, p-phenylenediamine, o-toluidine, o-benzidine and $N,N,N′,N′$-tetramethylbenzidine) are also used as solid state selective membranes for perchlorate ions [122]. A compact disk (~1 mm thick) of silver diethyldithiocarbamate containing a slight excess of silver nitrate functions as a selective membrane for the nitrate ion and shows a Nernstian behaviour over the nM to 0·1 M range of nitrate ions [123].

III. HETEROGENEOUS SOLID STATE
MEMBRANE ELECTRODES

Membranes of these types are prepared by dispersing ground dried powders of metal salts, chelates, ion-exchangers and macrocyclic compounds in inert matrices. The physical properties of these active materials (e.g. grain size, crystalline form, conditions of precipitation) are important to the functioning of the membrane. For example, when barium sulfate and silver chloride precipitates are used as active materials, they should be obtained by precipitation in the presence of excess sulfate and chloride ions respectively, whereas in the case of lanthanum fluoride and silver bromide the precipitation should be performed in the presence of excess lanthanum and silver ions respectively [124–127]. Secondary nucleation should be prevented by precipitation in the presence of p-ethoxychrysoidine. Papeschi et al. [127a] have demonstrated the advantages of membranes based on a silver sulfide matrix in which a silver halide is dispersed in very finely divided form.

The active materials are ground to a size of 1–15 μm and used, in general, in the ratio of about 1:1 with the inert matrix to give a membrane with the thickness of 0·2–1 mm. The particles of the active material should be in contact within the membrane in order to achieve conduction.

Heterogeneous membrane electrodes are summarized in Table 2.2. These electrodes are classified according to the matrix used and are discussed in the following sections.

A. Membranes with a silicone rubber matrix

At the ACHEMA symposium in 1964, Pungor called attention to a new type of electrochemical sensor based on precipitate-silicone rubber membranes [128]. The membranes are prepared [129] by dispersing the ground dried active material in silicone rubber monomer and polymerizing the mixture. The polymerization is conducted at low temperature in the presence of a suitable cross-linking agent (silan derivative) and a catalyst to ensure a certain degree of cross-linking. The degree of cross-linking affects the embedding of the particles in the membrane surface. The membranes in the form of disks (0·3–0·5 mm thick) are secured into the end of the glass tubes using silicone rubber adhesive [130] and the tube is filled with an electrolyte solution containing one of the ions present as an active material in the membrane. Subsequent conditioning of the membrane by soaking in the appropriate solution containing the ion of interest for a few hours is necessary [131]. Some Pungor-type electrodes are listed in Table 2.3.

Anion selective electrodes for chloride [132], bromide [124], and iodide [133] have been prepared using the corresponding silver salts as active materials [134, 135]. These membranes respond also to the cyanide ion

Table 2.2. Membrane types of some solid state heterogeneous selective electrodes

Ion measured	Membrane composition		References
	Active material	Matrix	
Alkyl benzene sulfonate	1,10-Phenanthroline iron(II)+ dioctylphthalate	Poly(vinyl chloride)	191
Ammonia	Potassium or calcium tetraphenyl borate	Poly(vinyl chloride)	197
Barium	Barium sulfate	Paraffin wax	263
	Barium sulfate	Cellophane	264
Bicarbonate	Anion exchange resin	Polystyrene	235
Bromide	Silver bromide	Silicone rubber	124
Cadmium	Cadmium sulfide + silver sulfide	Silicone rubber	142
	Cadmium sulfide + silver sulfide	Polyethylene	231
Caesium	Caesium dodecamolybdo phosphate	Silicone XA	150
	Caesium tungstoarsenate	Araldite	245
Calcium	Calcium stearate	Paraffin	255
	Calcium oxalate	Paraffin	252–254
	Calcium tributylphosphate	Poly(vinyl chloride)	161
	Calcium didecylphosphonium salt + dioctylphenyl phosphate	Poly(vinyl chloride)	162
	Calcium dihydrogen tetra(diphenyl-phosphate)+ dioctylphenyl- phosphate	Poly(vinyl chloride)	160
	Calcium didecylphosphonium + octylphenylphosphate	Graphite	205
	{NN'-bis-[11-(ethoxy carbonyl) Undecyl]-NN',3,4-tetramethyl-2,5-dioxahexane 1,6-dicarboxamide}+ sodium tetraphenyl borate	Poly(vinyl chloride)	166
	bis-(4-octylphenyl) hydrogen phosphate	Poly(vinyl chloride)	167

Table 2.2. (cont.)

Ion measured	Membrane composition		References
	Active material	Matrix	
	bis-[4-(1,1,3,3-tetramethylbutyl)phenyl]-phosphate	Poly(vinyl chloride)	168
	Calcium bis[bis-(2-ethyl hexyl)]phosphate	Poly(isobutyl vinyl ether)	243
Chloride	Silver chloride	Silicone rubber	124
	Aliquat 336	Graphite	205
	Silver chloride–silver sulfide	Polyethylene	233
Chromate	Barium chromate	Silicone rubber	139
Copper	Copper sulfide + silver sulfide	Silicone rubber	143–145
	Copper dithiazonate	Graphite	206, 207
	Cuprous iodide + dioctylphthalate	Poly(vinyl chloride)	170
	Copper sulfide + silver sulfide	Polyethylene	232
	Chlorocuprate(II)	Poly(vinyl chloride)	202
	Copper salicylaldoxime	Graphite	208
	Copper diphenyl thiocarbazone	Graphite	209
	Dowex 50w-XB + copper-2,2'-bipyridyl	Polystyrene	238
Fluoride	Lanthanum, calcium, thorium or samarium fluoride	Silicone rubber	126, 137
	Calcium fluoride	Poly(methylphenyl siloxane)	250
Halide	Silver halide	Thermoplastic resin	240, 241
	Silver halide + silver sulfide	Graphite + paraffin wax	203
	Silver halide + silver sulfide	Graphite	210
Iodide	Silver iodide	Silicone rubber	133
	Tetraoctylphosphonium salt + o-, or m-terphenyls	Graphite	211
Iron (III)	Tetrachloro ferrate (II) of Aliquat 336S	Poly(vinyl chloride)	198
Lead	Lead dithiazonate	Poly(vinyl chloride)	171
	Lead diethyl-dithiocarbamate	Poly(vinyl chloride)	174

Table 2.2. (cont.)

| Ion measured | Membrane composition | | References |
	Active material	Matrix	
Lithium	[Tris-(2-ethyl hexyl)] phosphate + Li(NN-diheptyl-NN,5,5, tetra methyl-3,7-dioxanonane-diamide	Poly(vinyl chloride)	182
Mercury (II)	Mercury (II) diphenyl thiocarbazone	Graphite	209
	Chloromercurate (II)	Poly(vinyl chloride)	201
Mono valent cations	Zirconyl phosphate	Polyethylene	234
Nitrate	Aliquat 336 (nitrate form)	Graphite	205
	Corning or Orion nitrate liquid exchanger	Poly(vinyl chloride)	184, 185
	Tetradecyl ammonium nitrate + dioctylphthalate	Poly(vinyl chloride)	183
	Liquid nitrate exchanger	Ceresine wax + graphite powder	204
	Tetraphenyl phosphonium nitrate	Glass frit	269
	Gentian violet nitrate	Glass frit	270
Organic anions and cations	Ion associate complex of Aliquat 336S with the organic ion	Poly(vinyl chloride)	192
	Tetrabutyl ammonium radical	Poly(vinyl chloride)	190
Perchlorate	Benzylhexadecyl dimethyl ammonium perchlorate	Graphite	213
	Liquid ion exchanger of Orion or Corning electrodes	Poly(vinyl chloride)	185, 186
	Tris-(4,7-diphenyl-1,10-phenanthrolinato) iron (II) perchlorate	Poly(vinyl chloride)	188
	Methylene blue-tetra fluoroborate	Glass frit	271
Phosphate	Aluminium, bismuth or magnesium phosphate	Silicone rubber	138, 140

Table 2.2. (cont.)

Ion measured	Membrane composition		References
	Active material	Matrix	
Potassium	Potassium(p-chlorophenyl)borate	Poly(vinyl chloride)	174, 175
	Valinomycin	Poly(vinyl chloride)	176, 177, 194
	Valinomycin	Silicone rubber	150
	Valinomycin	Poly (dimethyl siloxane)-poly(bisphenyl A carbonate)	251
	Crown compounds	Poly(vinyl chloride)	178–180
	Potassium zinc ferrocyanide	Silicone rubber	151, 152
	Stannic molybdate	Polystyrene	237
Silver	Silver sulfide	Silicone rubber	141
	Silver sulfide	Polyethylene	230
	Silver sulfide	Urea-glutaraldehyde copolymer	239
Sodium	Amberlite (IR-120)	Polystyrene	236
Sulfate	Barium sulfate	Silicone rubber	138
	Barium sulfate	Poly(vinyl chloride)	189
Thallium	Thallium dodecamolybdophosphate	Epoxy resin	244
	Thallium tungstoarsenate	Araldite	245
Thiocyanate	Silver thiocyanate	Thermoplastic resin	242
Toluene-p-sulfonate	Toluene-p-sulfonate	Graphite	213, 214
Zinc	Chlorozincate (II)	Poly(vinyl chloride)	199

and can be used for cyanide measurement [136]. Salts other than silver are used in silicone rubber matrix as active materials. Lanthanum fluoride, calcium fluoride and thorium fluoride impregnated in silicone rubber respond to the fluoride ion [126] over the range of pF 2–4. Incorporation of carbon with the precipitate has extended the response to pF 5. Samarium (III) fluoride is more effective for fluoride measurement and less subject to interference by other ions [137]. However, the heterogeneous fluoride membranes tend to lose their sensitivity after a few weeks and their performance is greatly inferior to the single crystal fluoride membrane. Sulfate [138] and chromate [139] membranes are prepared by impregnating

Table 2.3. Types and characteristics of some solid state heterogeneous membrane (Pungor type) electrodes[a]

Electrode	Model No.	Lower concentration limit	pH range	Interferences
Bromide	OP-Br-711	10^{-6}	2–12	S^{--}, I^-, CN^-
Chloride	OP-Cl-711	10^{-5}	2–12	S^{--}, I^-, Br^-, CN^-
Cyanide	OP-CN-711	10^{-5}	10·5	S^{--}
Iodide	OP-I-711	10^{-7}	2–12	S^{--}, CN^-
Perchlorate	OP-ClO$_4$-711	—	—	I^-
Sulfide	OP-S-711	10^{-17}	—	CN^-

[a] Radelkis Electrochemical Instruments, Budapest 62, Hungary.

barium sulfate and barium chromate, respectively, in silicone rubber matrix. Phosphate membranes are similarly prepared using bismuth phosphate [138], aluminium, chromium, cobalt, ferric, lanthanum, and magnesium phosphates and Dowex 1-X8 in the phosphate form [140].

Cation-selective electrodes for silver [141], cadmium [142], copper [143–145] and lead [146] ions are prepared by incorporating silver sulfide, cadmium sulfide–silver sulfide, copper sulfide–silver sulfide and lead sulfide, respectively in silicone rubber matrix. Silicone rubber membranes containing anhydrous and hydrated divalent transition metal phosphates of cobalt, nickel, copper and manganese proved not to be selective, since they respond to both mono- and divalent cations [131]. Zinc, nickel and aluminium complexonite (ion-exchanger resin containing iminodiacetic acid groups) impregnated in silicone rubber show selectivity towards ion valence type only and not to individual ions [147]. Cast membranes containing both cation and anion exchangers or a sandwich as means of eliminating the junction potential have been suggested [148].

Membranes selective to alkali metal ions have also been described. Caesium membrane [149] is prepared from caesium dodecamolybdophosphate as an active material and G.E. silicone XA 12X80 as a matrix in a 1:1 ratio. A silicone rubber–valinomycin membrane is described for potassium measurement [150]. A membrane which is selective to potassium ion over sodium and lithium but not over caesium, rubidium or ammonium ions is prepared from potassium zinc ferrocyanide (K_2 Zn_3 [Fe(CN)$_6$]$_2$), silastomer 70 and catalyst BC in the ratio of 4:4:1 [151, 152].

The response and applications of the silicone rubber membrane electrodes in organic solvents have been investigated. These electrodes do not respond in anhydrous solvents but they can be used in solutions of dimethyl-formamide, acetone and alcohols [153, 154] containing 10–40% water.

B. Membranes with a poly(vinyl chloride) matrix

The use of poly(vinyl chloride) as a matrix for the preparation of many membrane-type electrodes has been reported [155–157]. The active materials of the membrane are mixed with a solution of poly(vinyl chloride) in tetrahydrofuran or cyclohexanone solvent in the ratio of 1:2 to 1:10 and pressed at room temperature to give membranes with a thickness of about 0·3–0·5 mm. Hot pressing may be used in some cases [158]. The membranes are carefully mounted on a polished end of glass or poly(vinyl chloride) tubing filled with an electrolyte solution containing the ion of interest and an internal silver/silver chloride electrode (Fig. 2.6). The transport of mono- and divalent cations and monovalent anions through poly(vinyl chloride) membrane has been investigated and correlation between the selectivity in the transport (permeability) and the ion selectivity observed potentiometrically has been established [159]. Electrodes without internal reference solutions or platinum electrodes coated with the membrane materials have also been described.

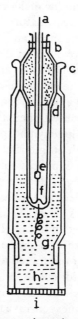

Fig. 2.6. Poly(vinyl chloride) membrane electrode. a, cable; b, Quickfit female socket SRP 10/19; c, silicone rubber adhesive; d, Quickfit male cone MF 15/0; e, solder joint; f, Araldite seal; g, silver–silver chloride electrode; h, internal reference solution; i, poly(vinyl chloride) ion-exchanger membrane. (Courtesy of the Division of Chemical Education, American Chemical Society.)

Cation membranes selective to alkaline earth metal ions have been described. Membranes selective to calcium ions are prepared from poly(vinyl chloride) with dioctylphenylphosphonate, calcium dihydrogen tetra(didecylphosphate) [160], tributylphosphate [161], nitrated (octylphenyl) phosphate [161a] and calcium didecylphosphate with di-n-octylphenylphosphonate [162], and tributylphosphate with 2-thenoyl trifluoro acetone [163, 164]. A study of several alkanols, using dioctylphenylphosphonate and dinonylphthalate as solvent materials indicated that the best electrodes for calcium are obtained with the use of dioctylphenylphosphonate [165]. A membrane prepared by mixing 0·9% {NN'-bis-[11-(ethoxycarbonyl)undecyl]-NN',3,4-tetramethyl-2,5-dioxahexane-1,6-dicarboxamide}, 64·3% of 2-nitrophenyl octyl ether, 34·4% of poly(vinyl chloride) and 0·4% sodium tetraphenylborate shows good response for the calcium ion [166]. Micro-electrodes for calcium measurement can be prepared by dissolving the calcium complex of bis-(4-octylphenyl) hydrogen phosphate in dioctylphenylphosphonate and a small amount of the product is mixed with a solution of poly(vinyl chloride) in tetrahydrofuran [167]. Then a boro-silicate glass tubing (1·8 mm o.d.) is drawn out, and its tip is broken to give a diameter of 10–20 μm. The glass is then siliconized in an atmosphere saturated with dimethyl dichlorosilane, the tip is left overnight in a droplet of the ion-exchange mixture, and the solvent is allowed to evaporate, leaving a coat on the inner and outer surface of the glass tube. A micro-electrode with a tip of 1 μm in diameter, suitable for measuring calcium in muscle cells (diameter 100 μm) and large ganglion cells (500 μm) has also been devised [168]. The electrode consists of a glass micro-pipette plugged by an inert poly(vinyl chloride) matrix containing the calcium salt of bis [4-(1,1,3,3-tetramethylbutyl) phenyl] phosphate and a mediator of low dielectric constant (dioctylphenylphosphonate). The electrode is filled with 0·1 M CaCl$_2$ solution with an internal Ag/AgCl reference electrode. This assembly is suitable for sample volumes down to 10 nl and has a rectilinear slope of 20–23 mV per decade of calcium concentration in the μM to 0·1 M range. Another electrode is based on the complexes of calcium and sodium tetraphenylborates [168a].

Membranes selective to transition metal ions have also been prepared. Finely divided ion-exchange resin (Acropore SA and SB) in acrylonitrile-poly(vinyl chloride) copolymer reinforced with nylon is used for the preparation of membranes sensitive for cobalt, ferric, zinc, copper and nickel ions [169]. Metal salts or complexes are also used in poly(vinyl chloride) matrix for the preparation of membranes sensitive to metal ions [170, 171]. Copper [170] and lead [171] electrodes are prepared from cuprous iodide using di-n-octylphthalate plasticizer and lead dithiazonate using diphenylphthalate plasticizer, or lead diethyldithiocarbamate using tetrachloroethane solvent [172], respectively. Metal complexes of the type Mx[N(II)L$_4$] where

L is SCN^- or I^-, M is Ag^+, Cu^+, Cu^{2+}, Pb^{3+}, Hg^{2+} or malachite green and N is Hg, Zn, Co or Ni, have been used as active material in a poly(vinyl chloride) matrix for the preparation of some metal selective membranes [173]. With $Pb\,[Hg\,(CNS)_4]$ membrane, for example, the electrode responds to Hg^{2+} and CNS^- only at concentrations > 10 mM with a change of 15 mV/decade for both ions. The internal solution used in these electrodes is $0\cdot1$ M chloride, iodide or thiocyanate of the metal being determined.

Membranes selective to alkali metal ions have been described. Potassium (p-chlorophenyl)borate ion exchanger incorporated in poly(vinyl chloride) matrix is used for the measurement of potassium ions [174, 175]. Macrocyclic and cyclic polyether compounds (Crown compounds) are used as active components in a poly(vinyl chloride) matrix for the preparation of potassium selective electrodes. Such electrodes are prepared by dissolving poly(vinyl chloride) and dibutylphthalate (1:3) in cyclohexanone and drifting the solution with valinomycin ($\sim 0\cdot1$ mg) followed by evaporation of the solvent to obtain a disk of ~ 1 mm thickness [176, 177]. Cyclic polyether compounds with six oxygen-binding atoms in an 18-membered ring and 30-membered ring compounds with ten oxygen atoms [178–180] are used in a similar manner instead of valinomycin. These membranes show good selectivity for potassium in the presence of sodium ($K_{Na/K}$ 360–450). However, both valinomycin and dimethyldibenzo-30-Crown-10 have been utilized for the preparation of membranes selective to potassium ions and suitable for the preparation of micro-electrodes [181]. Lithium ion-selective electrodes have been devised [182] by using tris-(2-ethylhexyl) phosphate in poly(vinyl chloride) matrix containing $5\cdot8\%$ of lithium ligand of (NN'-diheptyl-NN, 5,5-tetramethyl-3,7-dioxanonanediamide). The response of the electrode is rectilinear for the range $0\cdot1$ mM to M lithium and the selectivities of Li^+ over Na^+, K^+, Mg^{2+} and Ca^{2+} are 20, 140, 6000 and 1900, respectively.

Anion selective membranes based on the use of poly(vinyl chloride) matrix have also been described. A membrane selective for the nitrate [183] is prepared by incorporating tetradecyl ammonium nitrate in poly(vinyl chloride) using di-n-octylphthalate as plasticizer in the ratio of 1:30:10. Nitrate [184, 185], perchlorate [185, 186], chlorate [187] and halide [185] membranes are prepared using the liquid exchange materials used in the commercially available electrodes. A membrane selective for perchlorate ions can be prepared [188] by mixing $0\cdot09$ g of tris-(4,7-diphenyl-1,10-phenanthrolinato) iron (II) perchlorate, $0\cdot15$ g of poly(vinyl chloride), $0\cdot1$ ml of dioctylphthalate and 6 ml of tetrahydrofuran. Sulfate selective membranes have been prepared [189] by mixing $1\cdot2$ g of freshly precipitated barium sulfate with $0\cdot8$ g of powdered poly(vinyl chloride) and $0\cdot24$ g of dioctylphthalate. The mixture is stirred in an agate mortar with addition of cyclohexanone, allowed to swell for one day and dried at 70°C for 3 days.

Organic cation and anion selective membranes are prepared from the tetrabutylammonium radical of the corresponding ion with poly(vinyl chloride) plasticized with N,N-dimethyloleamide (Hallcomid 18-OL) [190]. Membranes selective for alkyl benzene sulfonate anions are prepared from the 1,10-phenanthroline-iron (II) complex using di-n-octylphthalate plasticizer [191].

Selective electrodes with poly(vinyl chloride) matrix and without internal reference solutions have also been described for both anions and cations. These electrodes are prepared by coating a platinum wire with solutions of poly(vinyl chloride) and the active membrane species using suitable plasticizers and solvents. This type of electrode is known as a "coated wire electrode". Electrodes selective for perchlorate, halide, thiocyanate, sulfate, oxalate, acetate, benzoate and salicylate are prepared [192, 193] by dipping a fine platinum wire into a mixture of poly(vinyl chloride) in cyclohexanone with the ion associate complex solution of the corresponding acid salt of Aliquat 336S.

Coated wire electrodes selective for some cationic species have been reported. Potassium electrodes are similarly prepared using valinomycin and didecylphthalate as plasticizer [194]. The calcium electrode is prepared by coating a platinum wire with a mixture of poly(vinyl chloride) and calcium alkylphosphate in the presence of a suitable plasticizer [195]. A poly(vinyl chloride) membrane containing calcium exchanger, mounted to a tube, and connected to a graphite rod is used directly without an internal reference solution in the tube [196]. Platinum wire coated with a poly(vinyl chloride)-di-butylphthalate (35:62) membrane containing 3% sodium, potassium or calcium tetraphenylborate gave a nearly Nernstian response to ammonium ions in the range 10^{-1}–10^{-3} M and is suitable for potentiometric titration of ammonium ions with calcium tetraphenylborate [197]. Platinum wires coated with Aliquat 336S in the forms of tetrachloroferrate (II) [198], chlorozincate (III) [199], chlorocadmate (II) [200], chloromercurate (II) [201] and chlorocuprate (II) [202] with poly(vinyl chloride) in the ratio of 7:3 show response towards iron (III), chlorozincate (II), cadmium (II), mercury (II) and copper (II), respectively.

C. Membranes with a carbon or graphite matrix

Graphite or carbon in the form of paste or rods are used as matrices for the preparation of some ion-selective and redox electrodes. These electrodes have no "memory effect", because the test samples cannot be adsorbed on the surface of the electrodes.

Halides and silver selective membranes have been prepared from pastes of carbon–paraffin wax (in the ratio of 3:1) as matrices and silver halides–silver sulfide as active materials [203]. A nitrate electrode prepared by mixing Orion liquid nitrate exchanger (92-07-02) with 4:1 ceresin-wax and graphite powder has been described [204]. The paste is held in PTFE rod (10 mm diameter) with a hole (10 mm deep) at one end and the working surface is renewed by pressing out the used end by means of a stainless steel piston which also provides electrical contact. Chloride, nitrate and calcium electrodes are prepared by mixing graphite powder with a suitable ion-exchange material containing the ion to be determined. For chloride, nitrate and calcium Aliquat 336, Aliquat 336 in the nitrate form and calcium didecylphosphate in di-n-octylphenylphosphate are used, respectively [205].

Graphite rods impregnated or coated with metal dithiazonates in xylene [206, 207] have been used as sensors for the determination of some heavy metals such as silver, mercury, ferric, and copper. Copper electrodes can also be prepared by the impregnation of graphite in copper salicylaldoxime [208] or copper diphenylthiocarbazone [209]. Electrodes sensitive to halide ions are prepared by suspending molten silver halides and silver sulfide mixtures on a rough surface of a graphite rod [210] previously made water-repellent with PTFE and polished on a hot stainless steel surface at 100–200°C.

The iodide membrane electrode consists of a disk (1 cm diameter and 1 mm thick) prepared by compressing a mixture of graphite, PTFE and a solution of ~M of tetra-octylphosphonium nitrate [211] in a mixture of m- and o-terphenyls, and responds to iodide ions in molten alkali nitrates at temperatures up to 160°C. Selective electrodes prepared by supporting an organic liquid phase saturated with the ionic species to be determined or an organic solution containing an extractable ion associate complex or metal chelate on carbon have been reported. The selectivity of such electrodes is closely related to the extraction behavior of the system used. These electrodes can be applied to any ionic species for which a suitable extraction system is available [212]. PTFE tube (~6 mm i.d.) fitted at the lower end with a carbon or graphite plug and containing a nitrobenzene solution or benzylhexa-decyldimethylammonium nitrate, iodide, perchlorate or toluene p-sulphonate [213, 214] has also been described for the determination of organic sulfonate as well as inorganic anions.

The graphite electrodes can be employed in redox titration, acid-base, dead-stop measurements, and voltammetry [215]. The use of the graphite electrode in voltammetry has the advantage that the residual current caused by the adsorbed oxygen is very low and the electrodes are easy to handle since the surface of the electrode is water-repellent. The relation between the surface area and the current constants shows that electrodes with a very small surface area are to be preferred [216–218]. Silicone rubber-based electrodes containing graphite have also received some attention in redox potentiometric titrations and voltammetry [219–224]. The graphite electrodes can be used not only as solid state redox electrodes [225, 226], but also as pH sensors [227–229].

D. Membranes with miscellaneous matrices

Polymeric plastic matrices such as polyethylene, polystyrene, epoxy resins, thermoplastic resins, thiourea-glutaraldehyde copolymer as well as paraffin, cellophane and collodion have been used.

Polyethylene membranes selective to silver [230], cadmium [231], copper [232], and chloride [233] ions are prepared by hot pressing of polyethylene and silver sulfide, cadmium sulfide–silver sulfide, copper sulfide–silver sulfide, and silver chloride–silver sulfide, respectively. A synthetic cation exchanger with about 50% zirconyl phosphate bound into polyethylene is sensitive to monovalent cations [234]. Heijne et al. [234a] used mixed precipitates of lead and silver sulfides for lead-selective electrodes.

Polystyrene is used as a matrix for the preparation of bicarbonate, sodium and potassium electrodes. A membrane suitable for the measurement of the bicarbonate ion [235] in the range of 1–10 mM is prepared from anion exchange resins (e.g. AV-17, AN-18, ARA-12, AR AP-5, Dowex and Dowex 2-X4) in 50% polystyrene. However, these membranes do not exhibit selectivity towards the bicarbonate ion. Amberlite (IR-120) dispersed in polystyrene and soaked in sodium chloride solution shows a good response towards sodium ions [236]. Incorporation of the hydrogen form of stannic molybdate gel into 15–25% polystyrene binder gives membrane selectively responding to potassium [237] ion with a Nernstian behavior largely independent of the nature of the anion species. A copper membrane can be prepared from Dowex 50W-X8 resin impregnated with a copper-2,2'-bipyridyl complex incorporated into polystyrene [238].

A thiourea–glutaraldehyde copolymer matrix has been utilized for the preparation of the silver electrode [239] using silver sulfide as an active membrane material. Halide [240, 241] and thiocyanate [242] selective membranes prepared by dispersing finely powdered silver halide and silver thiocyanate, respectively, in thermoplastic resin and casting the resin into a thin disk have been described. Poly(isobutyl vinyl ether) membrane containing calcium bis[bis-(2-ethylhexyl) phosphate] as an active component can be used for calcium measurement [243].

A membrane prepared from thallium dodecamolybdophosphoric acid incorporated in epoxy resin shows selectivity towards thallium ions [244]. Disks consisting of a 1 : 1 mixture of caesium tungstoarsenate or thallium(I) tungstoarsenate and Araldite may be used [245] as membranes for caesium and thallium(I) ions, respectively. Metal-dithizonates impregnated plastic membrane [246–249] electrodes are extensively studied for a variety of titrations. A membrane selective for fluoride ions is prepared by mixing powdered calcium fluoride with a solution of poly(methylphenyl siloxane) resin in toluene [250]. Potassium electrodes constructed from a membrane of a poly(dimethyl siloxane) poly(bisphenyl A carbonate) block copolymer with some cyanoethyl substitution, into which a potassium valinomycin–tetraphenylborate complex has been incorporated shows Nernstian response without response change for three years [251].

Paraffin membranes incorporating calcium oxalate [252–254] or calcium stearate [255] and non-ionic detergent with a gauze on which the membrane is suspended have been described as early as 1957. A nitrobenzene solution of ion-pair formed from the Eriochrome black T anion and the zephiramine cation admixed with naphthalene when solidified on a platinized platinum electrode, gives a response to Eriochrome black T in the range of 0·01 to 1 mM at pH 9 and can be used in complexometric titration of metal ions with EDTA [256]. However, these membranes are porous and not specific [257–259]. They probably function through micro cracks in the paraffin membrane [260].

Beckman solid calcium electrodes (Model No. 34608) are prepared by dissolving calcium dioctyl or didecylphosphate (3–5 parts by weight) in an ether-alcohol solution of collodion, and the active organophosphate-collodion membrane remaining after solvent evaporation is used (Fig. 2.7). Antibiotics dissolved in nitrobenzene and supported on a cellulose ester membrane have been utilized as a sensor for the measurement of calcium ions [261]. Cellulose triacetate filter may also be used for the potassium

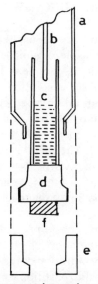

FIG. 2.7. Beckman solid ion-exchanger membrane electrode. a, electrode body; b, reference electrode; c, internal reference solution; d, O-ring for sealing the electrode membrane; e, sealing screw thread; f, electrode membrane.

membrane [262]. Barium sulfate–paraffin membranes [263] and barium sulfate–cellophane [264] membranes, prepared by placing barium hydroxide on one side of the cellophane membrane and sulfuric acid on the other side, have been described, but both are not perfectly selective. However, barium sulfate in parchment paper is used as an indicator electrode for aqueous and non-aqueous acid-base titrations [265–267]. Silver chloride, silver phosphate and silver tungstate incorporated in parchment paper are used for the potentiometric determination of fixed site densities and to measure solvent transfer [268].

A glass filter (G4 glass frit), when impregnated with organic solvents containing electro-active materials, acts as membrane specific to a particular ion. Impregnation with a solution of tetraphenylphosphonium bromide [269] in tetrachloroethane or nitrobenzene solution, previously mixed and stirred for 3 h with $NaNO_3$, or with a 0·1% solution of gentian violet [270] in tetrachloroethane, nitrobenzene, chloroform or chlorobenzene, previously mixed with M $NaNO_3$ form a nitrate sensitive membrane. Similarly, a perchlorate electrode can be constructed using methylene blue-tetrafluoroborate solution in tetrachloroethane [271]. Films of antimony-doped tin oxide on glass show a Nernstian pH response upon treatment with sodium hydroxide; after exposure to sulfur vapor, the electrodes resemble silver sulfide electrodes in sulfide response [271a].

IV. LIQUID MEMBRANE ELECTRODES

Liquid membrane selective electrodes are those in which the membranes are composed of water-immiscible organic solvents containing the ion of interest in the form of a salt or complex. The membrane is interposed between a standard and the test ion solutions. At the membrane interface, a process of ion exchange can take place between the ion of the ion-site salt in the organic phase, and the free ion in the aqueous phase. The suitability of the membrane depends on the characteristics of the liquid ion-exchanger materials used. The membranes should be insoluble in the aqueous phase, with reasonable viscosity to avoid loss by flow across the membrane, with a high degree of purity, selectivity, stability and ability of entering into a rapid mobile ion exchange equilibrium with the ion of interest.

The lower limit of useful electrode response depends mainly on: (i) the concentration of the ion in the aqueous phase of both the inner standard and test solutions; (ii) the concentration of the ion in the organic membrane; (iii) the solubility constant and partition coefficient of the complex ion between the aqueous and organic phases. Although the lower limit of response can be improved by either increasing the molecular weight of the site group, or by reducing the concentration of the salt in the organic phase, increasing the molecular weight beyond a certain limit may cause precipitation or gelling in the organic membrane phase. On the other hand, using a low concentration of the salt ion may lead to problems associated with high membrane resistance and long response time.

The e.m.f. of the liquid membrane electrodes can be described by equation 30.

$$E = E_0 - (RT/F) \ln \{0 \cdot 5[C + \sqrt{(C^2 + Ax)}]\} \qquad (30)$$

$$Ax = 4\sigma^2/bx \qquad (31)$$

where Ax is related to the limit of detection provided that the ion-exchanger dissociates completely in the membrane phase, σ represents the concentration of the ion-exchanger in the membrane and bx is related to the selectivity [272].

The liquid ion-selective electrodes are composed of an aqueous inner filling solution of the ion of interest ($\sim 0 \cdot 1$ M) and organic liquid exchanger materials which act as membranes. The electrodes are so constructed that the organic membrane separates the standard solution and the test ion solution. Various assemblies have been constructed to satisfy this requirement. In one of these assemblies, the liquid membrane is held inside a glass tube with a cellulose dialysis membrane at its end to prevent the loss of the organic liquid phase (Fig. 2.8). The internal standard aqueous solution

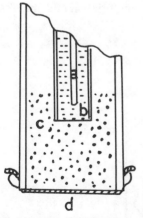

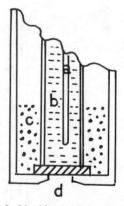

FIG. 2.8. Liquid membrane electrode (type A). a, internal reference electrode; b, agar gel with 2% of $0 \cdot 1$ M aqueous reference solution; c, liquid organic exchanger; d, electrode membrane (cellulose dialysis membrane or acetate filter disk of about 100 nm pore diameter).

FIG. 2.9. Liquid membrane electrode (type B). a, internal reference electrode; b, internal aqueous reference solution; c, liquid organic exchanger; d, electrode membrane (cellulose dialysis membrane or acetate filter disk of about 100 nm pore diameter).

is placed in a small inner tube filled with agar gel in which an inner silver/silver chloride reference electrode is immersed. These types of electrodes suffer from the high resistance and a relatively long response time due to the thick layer of the organic phase [273].

In another assembly (Fig. 2.9), the organic liquid is held in the pores of a thin disk of Millipore filter membrane of about 100 nm pore diameter. The inner chamber is filled with the internal aqueous solution in which the inner reference electrode is immersed. This avoids the use of a thick layer of the organic phase. Orion liquid membrane electrodes of the Series 92-type are three phase systems constructed on a similar basis (Fig. 2.10). The standard inner aqueous solution is separated from the outer aqueous test sample solution by a thin layer of immiscible organic exchanger phase held in the inert pores of the membrane. These electrodes should be maintained at a 20° angle to the vertical during operation to avoid trapping of air bubbles on the membrane.

Porous membranes soaked and saturated with the ion-exchanger material may also be used. Orion liquid membrane electrodes of Series 93-type contain such a type of membrane (Fig. 2.11). An electrode system which consists of a porous membrane soaked with 0.2 M thallium (I)-OO-didecyl phosphorodithionate in cyclohexane and a mM thallium (I) chloride, and 0.1 M potassium chloride in a saturated solution of silver chloride as an internal solution, has been described for the measurement of thallium (I) ions [274]. Electrodes selective for antimony [275] and gold [276] are similarly prepared by soaking a neutral-rubber membrane with a suitable solvent containing the liquid ion-exchanger material; liquid membranes

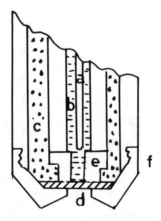

FIG. 2.10. Orion liquid membrane electrode (Series 92). a, internal reference electrode; b, internal aqueous reference solution; c, liquid organic exchanger; d, electrode membrane; e, membrane spacer; f, screw-cap. (Courtesy of Orion Research Inc.)

which consist of a poly(vinyl chloride) gel [277] containing the ion-exchanger substance or a gel layer of agar–agar [278] saturated with the ion-exchanger in the ratio of $1:3$ may be used (Fig. 2.12). Corning liquid membrane electrodes are constructed in such a manner that the organic phase is held in an inert ceramic plug or Pyrex frits (Fig. 2.13). The ceramic pores are siliconized to prevent wetting by the aqueous phase since a noisy response is observed if the pores are wet.

Other designs, not commercially available, but which may be of value in particular cases, have also been described. Micro-electrodes [279] of the type shown in Fig. 2.14 or of the double-barrelled type [280] (Fig. 1.29) may be used in biological and medical measurements. The ordinary macro-electrodes can be used in conjunction with a special micro dish [281] to permit the analysis of as low as 300 μl of sample solutions. A cell [282] as shown in Fig. 2.15 may be used for the measurement of the e.m.f. developed by the test solution provided that the organic liquid membrane phase is lighter than the aqueous solution (e.g. benzene solution of the ion exchanger). The cell is composed of two glass arms connected by a "$\cap$" shaped bridge. One arm is filled with a standard reference solution, the other is filled with the test solution (20 ml in each), and the bridge is filled with about two milliliters of the organic membrane phase. This assembly has the disadvantage of the need to renew all the solutions for each measurement and being incompatible with liquid membranes heavier than the aqueous phase.

The liquid exchanger membranes can be classified into three categories which are discussed below. A summary is given in Table 2.4. Some commercially available liquid membrane electrodes are listed in Table 2.5.

A. Cation liquid exchanger membranes

1. Phosphate Ester System
Ion-site salts derived from diesters of phosphonic acid are used as liquid exchanger membranes for calcium. Diesters with alkyl moiety in the range C_8–C_{16} form stable calcium complexes with good selectivity and without interference from monovalent cations. These complexes are used as liquid exchanger materials in suitable solvents [273, 283–285].

The use of calcium didecylphosphate dissolved in di-n-octyl phenylphosphonate was first described by Ross [273] and is probably used in the Orion calcium electrode (Model 92-20). The suitability of various calcium electrodes based on the use of metal salts of di(n-octyl phenyl)phosphoric acid, di(p-n-octyl-o-nitrophenyl) phosphoric acid and di(p-n-octyl-o-bromophenyl) phosphoric acid has been investigated [286]. The nature of the solvents in which the exchanger materials are dissolved is a very important factor governing the membrane selectivity. In solvents with highly polar substituent groups, such as di-n-octyl-phenylphosphonate [286a], good

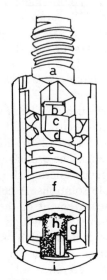

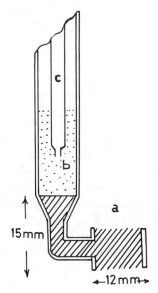

FIG. 2.11. Orion liquid membrane electrode (Series 93). a, module housing; b, electrical contact; c, internal reference element (Ag/AgCl); d, internal aqueous reference solution; e, spring; f, internal sensing assembly; g, porous plastic reservoir saturated with ion exchanger; h, organophilic porous membrane; i, ion-sensitive area. (Courtesy of Orion Research Inc.)

FIG. 2.12. Iodide ion-selective electrode with gel membrane (according to Novkirishka and Christova, *Anal. Chim. Acta* **78**, 63 (1975)). a, gel membrane (iodide form of Dowex 2X8, 20–50 mesh in agar-agar in the ratio 3:1); b, 0·1 M potassium chloride; c, saturated calomel electrode; d, glass electrode body. (Courtesy of Elsevier Scientific Publishing Co.)

selectivity for calcium relative to magnesium and other alkali earth ions is obtained, but in the presence of zinc, lead and ferric ions the selectivity is poor. With less polar solvents such as decanol, the membranes respond to all the alkaline earth ions without any selectivity for any of these ions. The solvent used should help minimize the solubility of the calcium phosphate ester in the aqueous phase.

An internal filling solution of calcium chloride (10^{-3} M) is used in the inner compartment of the electrode and contacts the inside surface of the membrane disk. A stable potential is developed by the calcium ion between the inside of the membrane and the filling solution, while a stable potential is established by the chloride ion between the silver/silver chloride internal reference electrode and the filling solution [287].

Ag; AgCl(s)	Internal reference solution of calcium (10^{-3} M Ca^{2+})	Calcium phosphate ester in the organic solvent (membrane)	Test calcium solution	AgCl; Ag

$$(32)$$

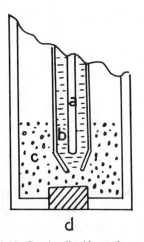

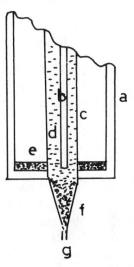

FIG. 2.13. Corning liquid membrane electrode. a, internal reference electrode; b, internal aqueous reference solution; c, liquid organic exchanger; d, porous siliconized ceramic plug.

FIG. 2.14. Orme's liquid membrane microelectrode (Reference 279). a, outer glass body (4 mm diameter); b, reference electrode; c, glass tube (0·4 mm diameter); d, internal aqueous reference solution; e, silicone rubber sealing cement; f, liquid organic ion exchanger; g, tip diameter about 5 μm. (Courtesy of J. Wiley & Sons Inc.)

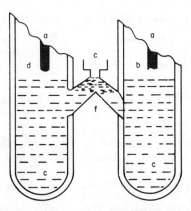

FIG. 2.15. Electro-analytical cell with a liquid membrane sensor (according to Scibona, Mantella and Danesi, *Anal. Chem.* **42**, 844 (1970)). a, reference electrode; b, internal aqueous reference solution; c, organic liquid ion exchanger; d, test solution. (Courtesy of The American Chemical Society.)

Table 2.4. Types of liquid membrane ion-selective electrodes

| Ion measured | Membrane composition | | References |
	Ion exchanger material	Solvent	
Acetate	Methyl tricaprylyl ammonium acetate (Aliquat 336S)	1-Decanol	315
Ammonium	Nonactin	Dibutylsebacate	346
	Nonactin (72%)+monactin (28%)	Tris (2-ethylhexyl) phosphate	346
Antimony	Antimony chloride-sevron red or flavinduline O	o-Dichlorobenzene	275
Barium	Barium Igepal C0800+ sodium tetraphenylborate	—	295
Benzoate	Methyltricaprylyl ammonium benzoate	1-Decanol	315
Bicarbonate	Tridecyl ammonium salt	Dibutyl phthalate+ chlorobenzene (1:1)	311
Boron tetrafluoride	Tris(substituted-o-phenanthroline)nickel	—	328
Bromide	Cetyltrimethyl ammonium bromide	Octanol	309
	Methyltricaprylyl ammonium bromide	1-Decanol	315
	Tetraheptyl ammonium bromide	Benzene	310
Caesium	Caesium-tetraphenyl borate	Nitrobenzene	294
Calcium	Calcium didecyl phosphate[a]	Dioctylphenyl phosphonate	373
	Nonylnaphthalene sulfonic acid	Nujol	335
Carbonate	Tricaprylylmethyl ammonium salt	Trifluoroacetyl p-butylbenzene	312
Chloride	Cetyltrimethyl ammonium chloride	Octanol	309
	Methyltricaprylyl ammonium chloride	1-Decanol	315
	Dimethyldistearyl ammonium chloride[b]	—	
	Tetraoctylphosphonium chloride	Decanol	322
	Tetraheptyl ammonium chloride	Benzene	310

Table 2.4. (cont.)

Ion measured	Membrane composition		References
	Ion exchanger material	Solvent	
Chromium	Dinonyl naphthalene sulfonic acid	Nujol	355
Copper	2'-hydroxy-3-ethyl-5'-methyl hexanophenone oxime copper	1-Decanol	297
	Dinonyl-naphthalene sulfonic acid copper salt	Nujol	335
	R—S—CH$_2$—COOCu	—	333
Formate	Methyl tricaprylyl ammonium formate	1-Decanol	315
Gold	Auric chloride-safranine O	o-Dichlorobenzene	276
	Tetraphenyl arsonium Au(CN)$_2$	Dichloroethane or chloroform	326
	Tetraphenylarsonium-[AuCl$_4$]$^-$	Dichloroethane	327
	1,1-Methylene bis-(4-ethyl-3,5-dipropylpyrazole)-[AuCl$_4$]$^-$	Nitrobenzene	339
Iodide	Cetyltrimethyl ammonium iodide	Octanol	309
	Methyltricaprylyl ammonium iodide	1-Decanol	315
	Tetrahexyl ammonium iodide	Dodecanol	310
	Tridodecylethyl ammonium iodide	Nitrobenzene	277
Lanthanum	Dinonyl naphthalene sulfonic acid lanthanum salt	Nujol	335
Lead	R—S—CH$_2$—COOPb	—	333
	OO-Dialkylphosphoro dithionate	—	334
Lithium	Lithium chloride	Decanol	366
Molybdenum	bis(tetraethyl ammonium) pentakis(thiocyanate) oxomolybdate(V)	Nitrobenzene + 1,2-dichloro-benzene (2:3)	321
Nickel	Dinonylnaphthalene sulfonic acid nickel salt	Nujol	335
Nitrate	Nitron nitrate	Benzyl alcohol	338
	Tetraoctyl or tetradecyl ammonium nitrate	Chlorobenzene	313
	Tris(substituted-o-phenanthroline)nickel^c	—	314

Table 2.4. (cont.)

Ion measured	Membrane composition		References
	Ion exchanger material	Solvent	
	Cetyltrimethyl ammonium nitrate	Octanol	309
	Tetraheptyl ammonium nitrate	Benzene	310
	Methyltricaprylyl ammonium nitrate	1-Decanol	315
	Tetraoctyl, or acetyl or iso-phosphonium nitrate	Bromoethane or 1-bromopropane or 1-chlorobutane or chlorobenzene	322
	Tridodecylhexadecyl ammonium nitrate^d	n-Octyl-o-nitro phenyl ether	
	Tetraoctyl phosphonium nitrate	Decanol	322
	Vitamin B$_{12}$	1-Decanol	356, 357
Oxalate	Methyltricaprylyl ammonium oxalate	1-Decanol	315
Palladium	Trilauryl ammonium or tetraheptyl ammonium salt of tetrachloro palladium(II)	Benzene	282
Perchlorate	Prodigiosin	1-Decanol	356, 357
	Vitamin B$_{12}$	1-Decanol	356, 357
	Azoviolene form of N-ethyl benzothiazole 2,2′-azine	1,2-dichloro-benzene or 2-chloroethyl ether	
	Tetrapropyl ammonium or tetrabutyl ammonium perchlorate	Dichloromethane	316
	Methyltricaprylyl ammonium perchlorate	1-Decanol	315
	Tris(4,7-diphenyl-1,10-phenanthroline) iron (II) or 1,10-phenanthroline iron (II) or tris 2,2′-bipyridyl iron (II)	Nitrobenzene	330
	Brilliant Green perchlorate	Chlorobenzene	320
	Tetrahexylammonium perchlorate	Dodecanol	317
	Methylene blue perchlorate	Nitrobenzene	340
Phosphate	Quaternary phosphonium salt of triphenyl tin dihydrogen phosphate	—	324

Table 2.4. (cont.)

Ion measured	Membrane composition		References
	Ion exchanger material	Solvent	
Potassium	Potassium tetrakis(4-chlorophenyl)borate	—	293
	Potassium tetraphenyl borate	2-Nitrotoluene	292
	Valinomycin^e	Diphenyl ether	341, 343, 344, 345, 353
	Nonactin	Diphenyl ether	341, 343
	Gramicidin	Dimethyl phthalate	346
	Crown compounds	Diphenyl ether	361–364
Propionate	Methyltricaprylyl ammonium propionate	1-Decanol	315
Salicylate	Tetraphenyl ammonium salicylate	Decanol	318
	Tricaprylylmethyl ammonium salicylate	Decanol	315
Selenium	3,3'-Diaminobenzidine	Hexane	298
Sodium	Monensin	Ethylhexyldiphenyl phosphate	346
Strontium	Strontium Igepal C0-880+ tetraphenyl borate	1-Ethyl-4-nitrobenzene	296
Sulfate	Methyltricaprylyl ammonium sulfate	1-Decanol	315
Sulfonate	Hexadecyltrimethyl ammonium dodecyl sulfonate	Nitrobenzene	336
	Sulfonate salts of Crystal Violet, Malachite Green, Methyl Violet, Fuchsine basic or Brilliant Green	Nitrobenzene or 1,2-dichloro-ethane or chloroform	319, 320
	Methyltricaprylyl ammonium p-toluene sulfonate	1-Decanol	315
Thallium	Thallium(I)-OO-didecyl phosphorodithioate	Cyclohexane	274
	Thallium chloride-phenazineduline {9-phenyldipyrido-[2,3-a:3',2'-c]phenazine-9-ium chloride}	o-Dichloro-benzene	275
Thiocyanate	Methyltricaprylyl ammonium thiocyanate	1-Decanol	315

Table 2.4. (cont.)

Ion measured	Membrane composition		References
	Ion exchanger material	Solvent	
Thorium	Dinonyl naphthalene sulfonic thorium salt	Nujol	335
Zinc	Trilauryl ammonium or tetraheptyl ammonium salt of tetrachloro zinc	Benzene	282

[a] Orion 92-20, [b] Orion 92-17, [c] Orion 92-07, [d] Corning 476134 and [e] Orion 92-19 ion exchanger materials.

The change in potential due to calcium ion concentration is given by equation 33.

$$E = E_0 + 2 \cdot 3RT/2F \log[a_{Ca^{2+}}] \tag{33}$$

The electrode responds only to the ionized calcium down to 10^{-4} M solutions and does not respond to calcium complexes such as citrate and polyphos-

Table 2.5. Characteristics of some commercially available liquid membrane electrodes

Electrode	Manu-facturer	Model	Lower concentration limit	pH range	Interfering ions
Boron tetra-fluoride	Beckman	39620	10^{-5}	2–12	ClO_4^-, I^-
	Orion	92-05	10^{-5}	2–12	I^-
Calcium	Beckman	39608	5×10^{-4}	3–11	H^+, Mg^{2+}, Ba^{2+},
	Corning	476041	10^{-5}	5–10	Cu^{2+}, Zn^{2+}
	Orion	92-20	10^{-5}	5–11	Zn^{2+}
Chloride	Corning	476131	10^{-5}	1–12	I^-, ClO_4^-, NO_3^-, Br^-
	Orion	92-17	10^{-5}	2–11	Br^-, I^-, NO_3^-, ClO_4^-, OH^-
Divalent (Ca^{2+}, Mg^{2+})	Beckman	39614	10^{-5}	6–11	Zn^{2+}, Ba^{2+}
	Corning	476235	10^{-5}	5–10	Ba^{2+}, Sr^{2+}, Ni^{2+}
	Orion	92-32	10^{-8}	5–11	Zn^{2+}, Fe^{2+}, Ni^{2+}, Cu^{2+}
Nitrate	Beckman	39618	10^{-5}	2–12	ClO_4^-, ClO_3^-, I^-
	Corning	476134	10^{-6}	2·5–10	ClO_4^-, I^-
	Orion	92-07	10^{-5}	2–12	I^-, ClO_3^-, ClO_4^-
Perchlorate	Beckman	39616	10^{-5}	2–12	ClO_3^-, I^-
	Orion	92-81	10^{-5}	4–11	OH^-

phate. Equation 33 is obeyed over a limited range when the internal and external solutions are not too different in concentration. A large concentration gradient of calcium across the membrane, either a test solution too concentrated for a dilute filling solution or a test solution too dilute for a concentrated filling solution, causes a deviation from the Nernstian equation. However, most of the liquid ion exchangers used in the commercial calcium electrodes display a Nernstian behavior over a wide range of calcium concentrations, even when the internal solution is 0·1 M and the external is 5 M.

The response of calcium phosphate exchanger in acidic medium [288] to selectivity isotherm, and the selectivity characteristic [289] of the liquid ion exchangers has been investigated [290]. Presence of organic bases in the reference solution causes an apparent fall in calcium activity except for pyridine, 2-picoline, 2,4,6-collidine and lutidine which cause an apparent increase [291].

2. Tetraphenylborate System

Potassium tetraphenylborate in 2-nitrotoluene solvent [292] and potassium tetrakis (4-chlorophenyl)borate [293] show good selectivity for potassium over sodium up to the ratio of 100 : 1 and the response is Nernstian down to 10 μM of potassium ion at the pH range of 3–11. A saturated solution of caesium tetraphenylborate in nitrobenzene has been used for the preparation of a liquid membrane selective for caesium with Nernstian response in the range of 0·1 M to 0·1 mM of caesium [294].

Ion-exchanger solution consisting of a complex prepared from Igepal CO-880 with barium chloride or strontium nitrate and sodium tetraphenylborate shows high selectivity for barium [295] and strontium [296], respectively. These membranes are prepared by dissolving 0·2–0·5 g of the metal-Igepal complex in 5 ml of 1-ethyl-4-nitrobenzene and is usable for the measurement of the metals in the range of 10 μM to 0·1 M at pH 2–10.

3. Amine and Oxime Systems

The complex of 2'-hydroxy-3-ethyl-5'-methyl hexanophenone oxime with copper in decanol is used as the liquid membrane for copper [297]. Potassium, zinc, nickel, ferrous and cadmium ions do not interfere with copper but ferric and aluminium do. A saturated solution of 3,3-diamino benzidine in hexanol shows good selectivity for selenium [297a] and can be used as a membrane for the determination of selenium in the range of ~0·1 mM to mM at pH 2·5.

Selective membranes for either $[SbCl_6]^-$ or $[TlCl_4]^-$ can be prepared [275] from o-dichlorobenzene solution of the $[SbCl_6]^-$ and $[TlCl_4]^-$ salts of Sevron Red L (C.I. Basic Red 17), Sevron Red GL (C.I. 11,085) and flavinduline O (C.I. 50,000), and the $[TlCl_4]^-$ salt of phenazinduline o-{9-phenyldipyrido[2,3-a: 3',2'-C]phenazin-9-ium chloride}. The membranes give an almost Nernstian response over the range 10 mM to μM for Tl^+ or $0 \cdot 1$ μM for Sb^{5+}. The construction of a potassium selective electrode containing $0 \cdot 0005$ to $0 \cdot 05\%$ potassium dipicrylaminate solution in nitrobenzene, with a porous hydrophobic support, has been reported [298].

B. Anion liquid exchanger membranes

1. *Quaternary Ammonium System*

Long chain alkyl and aryl ammonium salt solutions ($\sim 0 \cdot 1$ M) in low dielectric constant solvents are known to behave as a liquid anion exchanger suitable for the preparation of anion liquid membrane electrodes [299–307]. Methyl-, propyl-, butyl-, hexyl-, heptyl-, octyl-, decyl-, dodecyl-, capryl-, lauryl-, stearyl- and phenyl-ammonium salts of various anions have been used in suitable solvents as liquid membranes selective to chloride, bromide, iodide, nitrate, perchlorate, carbonate, bicarbonate, carboxylate and sulfonate ions. The selectivity of these membranes is affected by the nature of the solvent used. A comparison study [308] made with tetraheptyl-ammonium nitrate in pentanol and in benzene and its derivatives shows that the selectivity factors with benzene and its mono- and di-chloro derivatives are similar and are greater than those with pentanol or nitrobenzene.

Liquid membranes selective to inorganic anions have been reported. Halide membranes (chloride, bromide, iodide) are prepared using cetyltrimethyl ammonium halides in decanol [309] or tetraheptyl ammonium halides in benzene [310]. Bicarbonate [311] and carbonate [312] selective membranes are prepared from tridecyl ammonium and tricaprylylmethyl ammonium or tetraoctyl ammonium [312a] salts, respectively. A nitrate selective membrane based on the use of tetraoctyl or tetradecyl ammonium [313], methyltrioctyl ammonium [314], cetyltrimethyl ammonium [309], tetraheptyl ammonium [310] and tricapryl ammonium [315] nitrates were described. Perchlorate membranes are prepared using tetrapropyl ammonium-, tetrabutyl ammonium- [316], tricapryl ammonium- [315], and tetrahexyl ammonium- [317] perchlorates. A per-rhenate electrode has been prepared from tetradecyl ammonium per-rhenate (317a).

Liquid membranes selective to organic anions were also described and used for the determination of carboxylic, sulfonic and amino acid compounds. Acetate, formate, propionate, oxalate, benzoate and p-toluene sulfonate selective membranes were prepared from the corresponding methylcaprylyl ammonium salts (Aliquat 336) in decanol which are effectively and selectively used for the analysis of these organic anions

[315]. Salicylate liquid selective membranes prepared from tetraphenyl ammonium salicylate were also reported [318]. A picrate-selective electrode has been made with tetrapentyl ammonium picrate [318a]. Aromatic sulfonate liquid selective membranes are prepared from the sulfonate salts of triphenyl methane dyes such as Crystal Violet, Methyl Violet, Basic Fuchsine, Malachite Green [319] and Brilliant Green [320]. Triphenyl methane dyes are also used to prepare electrodes for chlorate [320a], perchlorate, thiocyanate, tetrafluoroborate, and nitrate [320b].

Although the exchanger membranes based on quaternary ammonium salts are used mainly for the preparation of anion selective electrodes, they can be used for the preparation of some cation selective membranes through their anionic complexes. The reaction of trilauryl ammonium chloride (TLAHCl) and tetraheptyl ammonium chloride (THACl) with zinc or palladium gives the complexes $(TLAH)_2ZnCl_4$, $(THA)_2ZnCl_4$ and $(TLAH)_2PdCl_4$ and $(THA)_2PdCl_4$. The benzene solutions of these complexes show marked selectivity to zinc and palladium ions respectively and are used as liquid membranes [282]. Molybdenum [321] selective membranes based on bis(tetraethyl ammonium) pentakis (thiocyanate) oxomolybdate in $2:3$ nitro-benzene-dichlorobenzene mixture have been reported.

2. Quaternary Phosphonium and Arsonium Systems

Tetraisopentyl-, tetraoctyl- and tetraacetyl-phosphonium nitrates in bromoethane, bromopropane, chlorobutane and chlorobenzene are used as liquid membranes for the nitrate ion. Among these salts tetraacetyl phosphonium nitrate shows the best selectivity and exhibits Nernstian response to nitrate concentration in the range of 10 μM to 0·1 M [322]. The liquid membrane electrodes based on the nitrate or chloride of tetra-octyl phosphonium ions in decanol or nitrobenzene solution can be used for the direct determination of various acids in admixtures [323]. The quaternary phosphonium salt of triphenyltindihydrogen phosphate can be used as an ion exchanger of good selectivity for $H_2PO_4^-$ ions [324]. A perchlorate electrode has been described [325] based on the use of the perchlorate salt of tetrakis (triphenyl phosphine) silver(I) as a liquid membrane, a 0·45 μM millipore filter as a porous membrane and an internal reference electrode incorporating aqueous agar gel (2 M in NaCl and M in $NaClO_4$) in contact with Ag/AgCl.

The tetraphenyl arsonium-$Au(CN)_2$ ion pair in chloroform or dichloroethane can be used as a membrane selective to gold in cyanide solution [326]. An electrode contains 5 mM of $KAu(CN)_2$ as the internal electrolyte exhibits Nernstian response for 3 μM to 30 mM $Au(CN)_2$ at pH 2·5 to 11·5. Large amounts of citrate, ferricyanide, ferrocyanide and cyanide ions do not interfere but $Ag[(CN)_2]^-$ ions do. The cation associate complex of tetraphenyl arsonium-$[AuCl_4]^-$ in dichloroethane can also be used as a membrane selective for gold [327].

3. Metal Phenanthroline System

Transition metal complexes of orthophenanthroline function as anion exchangers. The metal is non-labile and its exchange rate at the membrane interface is very slow relative to the rate of anion exchange. The positive charge in sites of these ligands is delocalized by the effect of the aromatic π bond system. The low charge density sites form, therefore, stable complexes with large anions such as perchlorate, nitrate, iodide, bromide, chloride, copper- and boron-tetrafluoride.

The selectivity coefficient of tris-(1,10-phenanthroline) iron (II), tris-(4,7-diphenyl-1,10-phenanthroline) iron (II) in nitrobenzene, chloroform or amyl alcohol is relatively independent of such a system and follows [328] the sequence of decreasing selectivity in the order of BF_6^-, ClO_4^-, SCN^-, $I^- \approx BF_4^-$, NO_3^-, Br^-, Cl^-.

A patent on electrodes sensitive to perchlorate, bromide, iodide, nitrate and chlorate using a tris-phenanthroline complex of nickel and cobalt has been issued [329]. Perchlorates of tris-(4,7-diphenyl-1,10-phenanthroline) iron (II), tris-(1,10-phenanthroline) iron (II) and tris-(2,2'-bipyridyl) iron (II) in nitrobenzene are used as liquid exchangers with high selectivity for the perchlorate ion [330]. A liquid membrane sensitive to copper has also been devised [331, 332].

4. Sulfur-containing Systems

Charged sites of some groups containing sulfur show selectivity towards certain ions which form either insoluble salts or stable complexes. Compounds of the type $R-S-CH_2COO^-$ form a chelate ring with lead and copper. Membranes based on this site group show good selectivity and sufficient stability [333]. Selectivities of liquid membranes based on the nickel, cadmium and lead salts of OO-dialkylphosphorodithionate have been investigated [334] and found to be decreased in the order: lead, cadmium and nickel.

Dinonylnaphthalenesulfonic acid (DNNS) has exchange properties similar to those of Dowex 50 and can be used for the preparation of a liquid membrane selective to tri- and tetravalent cations [335]. The great selectivity of electrodes based on DNNS is explained by solvent extraction considerations [335a]. Hexadecyl tri-methylammonium dodecyl sulfate [336] in nitrobenzene is used as a liquid membrane in electrodes selective to arylsulfonate compounds.

The perchlorate salt of the azaviolene form of N-ethyl-benzothiazole 2,2'-azine in 1,2-dichlorobenzene or 2-chloroethylether [337] is used as membrane selective to perchlorate ions.

5. Nitrogen Heterocyclic Systems

Solutions of nitron nitrate in aniline, chloroform and benzyl alcohol are suitable as membranes for the nitrate ion [338], but benzyl alcohol is the solvent to be recommended. Electrodes containing nitron nitrate are suitable for the determination of mM to M concentrations of the nitrate ion in the presence of large amounts of fluoride, or phosphate and lesser amounts of chloride, carbonate, bicarbonate, sulfate or acetate.

The 1,1-methylene bis-(4-ethyl-3,5-dipropylpyrazole)-[AuCl₄]⁻ complex in nitrobenzene is useful as a membrane selective for gold [339] with a Nernstian response for 30 μM to 1·3 M HAuCl₄. Other nitrogenous systems capable for the formation of ion-pair complexes with the perchlorate ion may be used for the preparation of a perchlorate ion-selective electrode [340].

C. Neutral macrocyclic membranes

1. Antibiotic Systems

A particular class of macrocyclic antibiotics including nonactin (formula I), gramicidin and valinomycin (II) has been reported since 1962 to possess

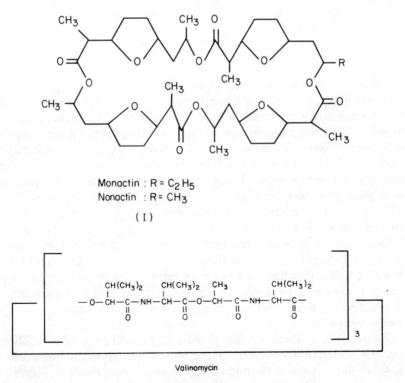

Monactin : R = C₂H₅
Nonactin : R = CH₃

(I)

Valinomycin

(II)

a significant effect on potassium transport through cell membranes. In 1967 these compounds were used as liquid exchanger membranes of higher selectivity to potassium over sodium [341]. The very great selectivity to potassium over sodium response (10^{-3}–10^{-4} M) has rendered these electrodes useful in situations where glass electrodes fail particularly in analyses of biological fluids [342].

Valinomycin [343] is dissolved in diphenyl ether and applied on a Millipore filter (MF type, Millipore Filter Corp., Bedford, Mass., U.S.A.). This liquid membrane sensor has a theoretical linear electrode function in the range 10^{-1} to 10^{-5} M with selectivity better than 10,000:1 for potassium over sodium [344] and electrode resistance of about 1 mΩ. The selectivity ratio for potassium against 15 different cations has been reported [345] indicating good performance towards potassium.

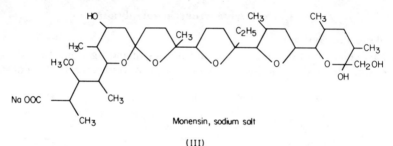

Monensin, sodium salt

(III)

Liquid membranes prepared from monensin (III) show a good response to potassium activity but the selectivity is inferior compared with valinomycin. Somewhat better results have been reported with nigericin, but electrodes made from it are inferior to the standard glass potassium electrode. This may be clear by comparing the selectivity for potassium using 0·04 M nonactin, 0·03 M nonactin and 0·009 M valinomycin in diphenyl ether as ion-selective components. These systems show selectivity for potassium over sodium in the ratio of 100, 200 and 4000:1, respectively [343].

Valinomycin or gramicidin in di-methylphthalate, monensin in ethylhexyl diphenylphosphate and nonactin in dibutyl serbacate can be used as liquid membranes selective to potassium, sodium and ammonium ions, respectively. A membrane prepared from 72% nonactin and 28% monactin in tris-(2-ethylhexyl) phosphate has been reported to sense ammonium ions in the presence of 10 times of potassium and 500 times of sodium [346]. Consequently, sensors for selective determination of ammonium ions have become possible.

These types of electrodes are prepared by holding the liquid exchanger materials in a porous cellulose acetate, filter paper, nylon and other supports [347]. Solid versions of the potassium and ammonium electrodes have been reported [348–350]. Micro-capillary [351] (Fig. 2.14) and double-barrelled

[280] (Fig. 1.29) electrodes for potassium have been developed. The latter is made from double-barrelled borosilicate glass tubing (3 mm o.d.), the tip being broken at a diameter of ~ 2 μm. The two channels are filled with $0 \cdot 1$ M KCl and $0 \cdot 1$ M NaCl respectively, and the ion-selective liquid (valinomycin and potassium-tetrakis (4-chlorophenyl) borate in di-octyl-phthalate) is drawn into the potassium chloride channel to a height of ~ 200 μm. Chlorinated silver wire is sealed into both channels. This electrode is superior to the conventional ion-exchange micro-electrode with respect to linear range, useful life (~ 6 weeks), selectivity for potassium over sodium, hydrogen, acetylcholine and tetramethyl ammonium ions, and the risk of intracellular contamination. However, the most serious interferences are caused by caesium and rubidium [352].

The evaluation and construction of some of the antibiotic alkali metal ion-selective electrodes have been reported [353]. Certain anomalies (non-Nernstian response) are observed in the behavior of these electrodes under certain conditions [354]. The relationship between the anomalous behavior and the composition of the electrode membrane has been discussed. A potassium selective valinomycin liquid membrane electrode has been shown to respond to positively charged surfactants and the interferences vary with the pH of the test solution [355].

Antibiotic compounds other than those reported above, and some bio-chemical compounds are reported to have selectivity toward some anions and can be used as liquid membranes [346]. The antibiotic prodigiosin [356, 357] in decanol shows a linear relation for perchlorate ions whereas vitamin B_{12} in decanol responds to nitrate and perchlorate ions in the range of 10^{-1} to 10^{-5} M.

Various theories have been proposed to explain the role of the antibiotics in ion transport through membranes. These mechanisms are based on the complex formation of these antibiotic compounds with the ion. The antibiotic molecule may act at the interface to enable the ion to pass into the membrane wherein they move as free ions. The permeability is enhanced by passing the ions through channels in the membrane. Consequently, the ions to be measured move through the membrane in the cavity of the antibiotic ligand which acts as a carrier or transport catalyst. It is possible therefore to assume that these compounds act as uncharged hydrophobic extracting species that develop interfacial and/or diffusion potentials across a hydrophobic liquid membrane creating electrodes highly selective to potassium, sodium and ammonium ions [358].

The e.m.f. of such sensors (e.g. with a potassium membrane) are given by the equation:

$$E = E_0 + RT/\text{F} \ln [a_{K^+} + \underbrace{U_{MS^+}/U_{KS^+} \times K_{M^+}/K_{K^+}}_{K^{Pot}_{K^+M^+}} a_{M^+}] \qquad (34)$$

Table 2.6. The selectivity constant of some antibiotics with alkali metal ions

Selectivity constant	M^+						
	Rb^+	K^+	Cs^+	NH_4^+	Na^+	Li^+	H^+
$K^{Pot}_{K^+M^+}$ for valinomycin in diphenylether	1·9	1	$3·8\times10^{-1}$	$1·2\times10^{-2}$	$2·6\times10^{-4}$	$2·1\times10^{-4}$	$5·6\times10^{-5}$
$K^{Pot}_{NH_4^+M^+}$ for (nonactin 72% + monactin 28%) in tris(2-ethyl-hexyl)phosphate	$4·3\times10^{-2}$	$1·2\times10^{-1}$	$4·8\times10^{-3}$	—	2×10^{-3}	$4·2\times10^{-3}$	$1·6\times10^{-2}$

where a_{M^+} is the activity of the interfering ion, U_{MS^+} and U_{KS^+} are mobilities of electrically charged complexes within the membrane, and K_{K^+} and K_{M^+} are equilibrium constants of the extraction process. Since in most cases $U_{MS^+} \approx U_{KS^+}$, the selectivity constant of liquid membrane with electrically neutral ligands is mainly given by the ratio of the equilibrium constants, which are determined in part by the complex formation constants of the antibiotics and also depend on the other components of the membrane. A correlation has been found between the selectivity in the ion transport (permeability) and the ion selectivity observed potentiometrically [159]. A low dielectric solvent favored the selectivity of the potassium membrane [359]. Table 2.6 lists the selectivity constants of some antibiotic membranes.

2. Cyclic Polyether System
Neutral polycyclic compounds other than the naturally occurring antibiotics have been studied in terms of their complexing ability and extraction properties towards alkali metal and alkaline earth ions in organic immiscible solvents [359–363].

The cyclic polyethers or "crown compounds" (see formula IV) synthesized in 1967 have gained attention for use as effective complexing agents in liquid membrane electrodes. Nearly theoretical response towards potassium is obtained by using these compounds in diphenyl ether solvent (about 10:1 in preference to potassium over sodium). Formation constants for some metallic complexes of rubidium and potassium with dicyclohexyl 18-crown-6, dibenzo-18-crown-6, benzo-15-crown-5 and dibenzo-30-crown-10 in a 1:1 solution of tetrahydrofuran and water show that the selectivity ratio for one ion over another is approximately equal to the corresponding ratio of complex formation constants [364].

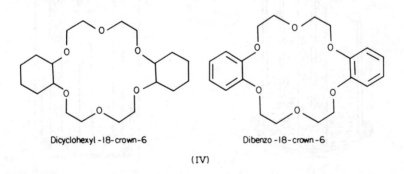

Dicyclohexyl-18-crown-6   Dibenzo-18-crown-6

(IV)

A liquid membrane selective for barium ion is prepared similarly using polyethyleneglycol (Igepal CO880) and shows a Nernstian response in the range of 10^{-1} to 10^{-5} M of the barium ion [365].

V. GAS-SENSING MEMBRANE PROBES

Gas-sensing membrane probes are highly selective potentiometric sensors used for the direct measurement of dissolved gases and ionic species that are easily converted into gases by pH adjustment. The term "probes" is used instead of "electrodes" because these sensors are complete electrochemical cells containing their own reference electrodes. These devices consist of a sensor electrode (e.g. glass or ion-selective electrode) located immediately behind a gas permeable membrane, whereby a very thin film of an internal electrolyte is sandwiched between this electrode and the gas-permeable membrane, and a second electrode acting as a reference electrode is located in the electrolyte [367] (Figs 2.16, 2.17, 2.18).

The high selectivity of these devices arises from the fact that only gaseous species pass through the membrane. Other advantages are also offered by using these probes. Thus:

(i) They are applicable to the analysis of ionic species for which no ion electrode is available. (For example, samples containing sulfite, nitrite, acetate and fluoride ions are converted in acidic media into sulfur dioxide, nitrogen oxides, acetic acid and hydrogen fluoride gases, respectively, and the ammonium ion is converted in alkaline medium into ammonia gas.)

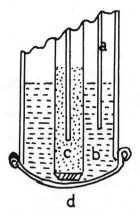

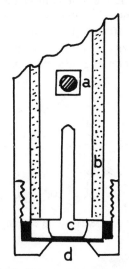

FIG. 2.16. Schematic diagram of a gas-sensing membrane probe: a, reference electrode; b, internal filling electrolyte solution; c, sensor electrode; d, gas permeable membrane.

FIG. 2.17. Orion gas-sensing membrane probe (series 95). a, reference electrode; b, internal electrolyte solution; c, sensor electrode; d, gas permeable membrane. (Courtesy of Orion Research Inc.)

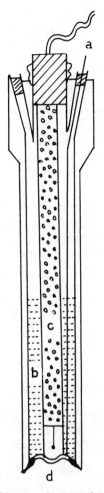

FIG. 2.18. EIL gas-sensing membrane probe. a, reference electrode; b, internal filling electrolyte solution; c, sensor electrode; d, gas permeable membrane. (Courtesy of The Chemical Society of London.)

(ii) They are free from liquid junction potential problems associated with the pH and the ion-selective electrodes, and are free from redox interferences.

(iii) Their potential response is Nernstian over a wide range of concentration permitting the application of the various measurement techniques and calibration procedures used with the ion-selective electrodes.

In 1957, the first gas-sensing membrane probe was developed for the measurement of carbon dioxide [368, 369]. The next probe was described for ammonia [370]. Since then, many efforts have been devoted to the development of various types of probes. Probes for carbon dioxide, ammonia, sulfur dioxide, nitrogen oxide, hydrogen cyanide, acetic acid and hydrogen fluoride are now commercially available.

A. Origin of probe potential

When the gas membrane probe is immersed in a solution containing the species to be measured, the gas (G) diffuses through the membrane depending on its concentration until the partial pressure of the gas is equal both in the thin electrolyte film and the sample:

$$G \text{ (sample)} \rightleftharpoons G \text{ (membrane)} \rightleftharpoons G \text{ (film)} \tag{35}$$

As a result of the gas dissolution in the electrolyte, the concentration of some ionic species in the electrolyte are changed. This change is detected by the sensor electrode [371, 372].

The equilibrium set up between the internal filling electrolyte (A^+ and B^-) and the gas (G) changes as a result of gas diffusion from the sample, causing an appreciable increase in the ionic species:

$$A^+ + B^- \rightleftharpoons G \text{ (gas)} \tag{36}$$

The presence of a sensor electrode selective to either A^+ or B^- will measure this change which is directly related to the concentration of the gas in the sample:

$$[A^+] = K[G]/[B^-] \quad \text{or} \quad [B^-] = K[G]/[A^+] \tag{37}$$

where K is the dissociation constant of the gas.

In general, the probe potential ideally changes as a function of the gas concentration according to the Nernst equation:

$$E = \text{constant} + RT/n\mathrm{F} \log [A^+] \tag{38}$$

$$E = \text{constant} \pm RT/n\mathrm{F} \log [G] \tag{39}$$

The sign of the last term is either positive for the acidic gases or negative for the basic gases. It is obvious that any parameter which affects the Henry's law constant of the system will affect the response of the probes since they sense partial pressure.

The response of many gas-sensing membrane probes is based on the change in the pH of the internal electrolyte film as a result of the gas diffusion process. This change is influenced by the equilibria established in the electrolyte system. This can be shown below with the carbon dioxide probe [373] as an example. Carbon dioxide diffuses and is dissolved in the aqueous solution of the internal thin electrolyte film to form carbonic acid:

$$CO_2(g) + H_2O \rightleftharpoons H_2CO_3 (Kp = a_{H_2CO_3/P_{CO_2}}) \tag{40}$$

$$H_2CO_3 \rightleftharpoons H^+ + HCO_3^- (K_{a_1} = a_{H^+} a_{HCO_3^-}/a_{H_2CO_3}) \tag{41}$$

$$HCO_3^- \rightleftharpoons H^+ + CO_3^= (K_{a_2} = a_{H^+} a_{CO_3^=}/a_{HCO_3^-}) \tag{42}$$

Because the ionic media in the sample and the film are constant, then:

$$[CO_2] \text{ sample} \approx K_D \cdot P_{CO_2}(\text{film}) \tag{43}$$

Because K_D is constant at a given temperature and the partial pressure of carbon dioxide gas at equilibrium is equal on either side of the membrane, and a glass electrode is used as a sensor, then the potential developed is given by the following equation:

$$E = E^0 \text{ glass} + RT/nF \ln a_{H^+} - E^0_{Ag/AgCl} + RT/nF \ln a_{Cl^-} \tag{44}$$

With the chloride concentration and activity being constant,

$$E = E' + RT/nF \ln a_{H^+} \tag{45}$$

Substitution for a_{H^+} (in equation 41), for $a_{H_2CO_3}$ (in equation 40) and for P_{CO_2} (in equation 43),

$$E = E' + RT/nF \ln K_{a_1} K_p K_D - RT/nF \ln a_{HCO_3} + RT/nF \ln [CO_2] \text{ sample} \tag{46}$$

This is simplified to the following equation:

$$E = E'' + RT/nF \ln [CO_2] \text{ sample} \tag{47}$$

The equilibria set up in the ammonia probe are relatively simple. Contrastingly, complex equilibria are established in the nitrogen oxide probe [372]. This is probably due to the participation of at least five species of nitrogen oxides (i.e. NO_2^-, NO_3^-, HNO_2, NO_2 and NO) and the various equilibrium constants with different temperature coefficients. The probe actually senses the equimolar mixture of NO and NO_2 gases in equilibrium with aqueous nitrous acid. It measures equally well the nitrite ion which is easily converted into nitrous acid by acidification:

$$NO_2^- + H^+ \rightleftharpoons HNO_2 \tag{48}$$

It is not known exactly what gaseous species pass from the acidic nitrite solution through the membrane. The term "nitrogen oxide probe NO_x" reflects this uncertainty.

B. Internal electrolyte solutions

The typical gas-sensing membrane probes based on the following cell compositions have been reported. Carbon dioxide sensing membrane probe [373]:

$$\text{Ag, AgCl} \left| \begin{array}{c} 0\cdot005 \text{ M NaHCO}_3 \\ 0\cdot02 \text{ M NaCl} \end{array} \right| \left| \begin{array}{l} 0\cdot005 \text{ M NaHCO}_3 \\ 0\cdot02 \text{ M NaCl} \\ \text{CO}_2 \text{ (aq.)} \end{array} \right| \text{glass electrode} \quad (49)$$

Nitrogen oxide sensing membrane probe [374]:

$$\text{Ag, AgBr} \left| \begin{array}{l} 0\cdot4 \text{ M KNO}_3 \\ 0\cdot1 \text{ M NaNO}_2 \\ 0\cdot1 \text{ M KBr} \\ \text{AgBr (S)} \end{array} \right| \left| \begin{array}{l} 0\cdot4 \text{ M KNO}_3 \\ 0\cdot1 \text{ M NaNO}_2 \\ 0\cdot1 \text{ M KBr} \\ \text{AgBr (S)} \\ \text{NO}_x\text{(aq.)} \end{array} \right| \text{glass electrode} \quad (50)$$

Ammonia sensing membrane probe [374–376]:

$$\text{Ag, AgCl} \left| \begin{array}{l} 0\cdot1 \text{ M NH}_4\text{Cl} \\ \text{AgCl (S)} \end{array} \right| \left| \begin{array}{l} 0\cdot1 \text{ M NH}_4\text{Cl} \\ \text{AgCl (S)} \\ \text{NH}_3 \text{ (aq.)} \end{array} \right| \text{glass electrode} \quad (51)$$

Ammonia sensors which comprise a silver/silver sulfide electrode and a solution of $0\cdot1$ M silver nitrate in $0\cdot1$ M ammonium nitrate as an internal electrolyte [377], or a graphite rod with silver sulfide or silver sulfide–copper sulfide rubbed into its tip and immersed in an electrolyte containing $0\cdot1$ M ammonium nitrate and $0\cdot01$ M sodium fluoride have been described. The response of these probes is 100 mV per decade of change in ammonia concentration. Similarly, probes for acetic acid [372] and sulfur dioxide [374] can be constructed using sodium acetate and sodium bisulfite or metabisulfite, respectively, as internal electrolyte solutions. Other gas sensing electrodes are given in Table 2.7.

The concentration of the electrolyte in the internal filling solution affects the time response and the potential stability of the probe. In general, more dilute electrolyte solutions give a faster time response than the concentrated solutions [372–374]. However, the concentration of sodium bicarbonate used as the internal electrolyte in the carbon dioxide probe has a different effect. At a low concentration of sodium bicarbonate, the sensitivity falls below the theoretical value, and the response time becomes very long [373]. Better precision is obtained using $0\cdot005$–$0\cdot01$ M of sodium bicarbonate and $0\cdot1$ M of ammonium chloride for carbon dioxide and ammonia gas probes, respectively [373, 374, 379]. In general, too dilute an electrolyte solution is not recommended for measuring high gas concentrations because

Table 2.7. Some gas-sensing electrode systems [372]

Measured species	Diffusing species	Equilibria in electrolyte	Sensing electrode
NH_3 or NH_4^+	NH_3	$NH_3 + H_2O \rightleftharpoons NH_4^+ + OH^-$	H^+
		$xNH_3 + M^{n+} \rightleftharpoons M(NH_3)_x^{n+}$	$M = Ag^+, Cd^{2+}, Cu^{2+}$
SO_2, H_2SO_3 or $SO_3^=$	SO_2	$SO_2 + H_2O \rightleftharpoons H^+ + HSO_3^-$	H^+
NO_2^-, NO_2	$NO_2 + NO$	$2NO_2 + H_2O \rightleftharpoons NO_3^- + NO_2^- + 2H^+$	H^+, NO_3^-
$S^=$, HS^-, H_2S	H_2S	$H_2S + H_2O \rightleftharpoons HS^- + H^+$	S^{2-}
CN^-, HCN	HCN	$Ag(CN_2)^- \rightleftharpoons Ag^+ + 2CN^-$	Ag^+
F^-, HF	HF	$HF \rightleftharpoons H^+ + F^-$	F^-
		$FeF_x^{2-x} \rightleftharpoons FeF_y^{3-y} + (x-y)F^-$	$Pt(redox)$
$HOAc$, OAc^-	$HOAc$	$HOAc \rightleftharpoons H^+ + OAc^-$	H^+
Cl_2, OCl^-, Cl^-	Cl_2	$Cl_2 + H_2O \rightleftharpoons 2H^+ + ClO^- + Cl^-$	H^+, Cl^-
CO_2, H_2CO_3, HCO_3^-, $CO_3^=$	CO_2	$CO_2 + H_2O \rightleftharpoons H^+ + HCO_3^-$	H^+
X_2, OX^-, X^-	X_2	$X_2 + H_2O \rightleftharpoons 2H^+ + XO^- + X^-$	$X = I^-$, Br^-

in such cases the response is non-Nernstian. Failure of existing models to predict pCO_2 electrode response times has been investigated [379a].

Variation of the concentration of the dissolved species on both sides of the membrane causes a transfer of water vapor across the membrane by osmotic pressure until the water activity is the same on both sides. This effect is more pronounced when the temperature varies between the sample and the thin electrolyte film [372, 374]. The transfer of water across the membrane will cause dilution or concentration of the electrolyte in the film and render the probe potential to drift. This drift depends on the rate of change in concentration in the film by interchange with the bulk of the internal electrolyte. This drift is not observed with ammonia and sulfur dioxide probes when the internal electrolyte solution is stronger than the sample solution [372]. However, with samples of higher electrolyte concentration than the internal electrolyte of the ammonia probe, a potential drift is noticed depending on the permeability of the membrane to water and the osmotic pressure gradient across the membranes.

The drift due to osmotic effects can be eliminated in the ammonia and nitrogen oxide probes by adding sufficient inert electrolyte to the internal filling solution [374]. This procedure is necessary for the determination of ammonia in concentrated salt solutions, or after Kjeldahl digestion [380, 381]. Gas permeable membranes insensitive to osmotic pressure may be used in the sulfur dioxide probe to avoid osmotic effects, but this may reduce the speed of response.

C. Gas permeable membranes

The membranes used in the gas probes are plastic films permeable to the diffusible gaseous species. The diffusion coefficient of the gases through the membrane, their partition coefficient between the membrane and the aqueous sample, and the membrane type and thickness affect the response time of the probes. Polyethylene and poly(vinyl fluoride) have been used for carbon dioxide gas [372, 373], as they have low permeabilities towards many other gaseous species. Homogeneous membranes made from silicone rubber and 0·025 mm thick are utilized in the sulfur dioxide and carbon dioxide probes. However, such membranes are easily damaged during assemblage or in the presence of sulfuric acid [373].

"Air gap" membranes consisting of thin microporous plastic films in which the transfer of the gas takes place through the air in the pores of the membrane have been advantageously used. The matrix material is hydrophobic and the membrane itself is essentially a thin air gap separating the sample and the internal electrolyte solutions. For most diffusible species, the permeability of the "air gap" membranes are about one thousand times greater than dense plastic films. Membranes with 0·1 and 0·001 mm thick PTFE are used for ammonia and carbon dioxide probes, respectively, and a polypropylene membrane of 0·025 mm thickness is used for the nitrogen oxide probe [373, 374, 382].

Some gas membrane probes are used with a liner between the membrane and the sensor electrode. For example, the Radiometer carbon dioxide probe (E 5036) has a layer of tissue paper, as a liner, in order to stabilize the thin film of electrolyte. However, the probes could operate equally well without the liner, according to Midgley [373].

D. Analytical techniques and applications

Measurements of gas species in their solutions can be made in either open or closed systems depending on the volatility of the gas. Since the volatility of ammonia is relatively low, it can be measured in an open beaker, whereas sulfur dioxide and nitrogen oxides should be measured in a closed system to avoid loss by volatility. This can be done by immersing the probe in the sample solution contained in a flask and the probe is sealed to the flask neck by an O-ring [374].

When the gas solution is non-aqueous or contains a surfactant which wets the membrane and allows liquid to penetrate, the probe cannot be immersed in these media [383]. In such cases, the sample solution is adjusted to the suitable pH, and is placed in a conical flask with a magnetic stirring bar (Fig. 2.19). The neck of the flask is fitted with a rubber stopper with a hole through which the electrode is fixed. This closed flask forms an air-tight system whose gas phase is saturated with water vapor and has a

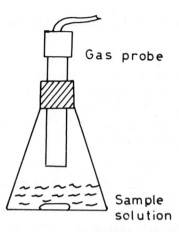

FIG. 2.19. Analysis of gaseous species by the gas-sensing membrane probe in a closed flask.

partial pressure of the gas in equilibrium with the solution. The electrode senses the partial pressure of the gas as it would when immersed in the gas solution, but with longer response time. Sulfur dioxide in wine is determined by adding sodium hydroxide solution to the samples in a closed vessel, followed by hydrochloric acid and TISB, and measurement of the liberated sulfur dioxide [384].

Gas measurements in flow systems (Fig. 2.20) using gas-sensing membrane probes have been described [374, 380, 385]. The electrode is attached to a flow-through cap, through which the gas solution flows and the samples are fed by an Auto-analyzer sampler and proportionating pump. This permits the analysis of 60 samples of ammonia per hour or 30 samples of sulfur dioxide per hour.

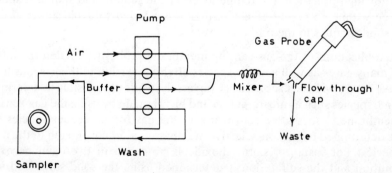

FIG. 2.20. Analysis of gaseous species by continuous flow technique (according to Bailey and Riley, *Analyst* **100**, 145 (1975)). (Courtesy of The Chemical Society of London.)

The following points may be considered for gas measurement by gas probes [373, 374]:

(i) The pH of the sample solution is adjusted in order to convert virtually all the sample solution into the gaseous phase. For the determination of ammonia, sulfur dioxide, nitrogen oxides, carbon dioxide and acetic acid, the pH's are adjusted to 12, 0·7, 1–7, 5–7 and 3, respectively. The reagents used for adjusting the pH should contain non-volatile species such as sulfuric and perchloric acids.

(ii) The presence of metals that are prone to form complexes with the gas should be masked. For example, determination of ammonia in the presence of zinc and copper ions can be performed using a buffer containing sodium hydroxide and EDTA, which not only masks the metals but also prevents precipitation of metal hydroxides that may absorb ammonia gas.

(iii) The quantitative release of gaseous species by pretreatment with a suitable reagent before adjustment may be needed. For example, sulfur dioxide in food and beverages is partly present as an aldehyde bisulfite addition product, which is not decomposed by the acidic buffer used for sulfur dioxide measurements. By alkali treatment at $pH > 12.5$, these compounds are rapidly decomposed to alkali bisulfite which upon acidification to pH 0·7 liberate SO_2 gas.

(iv) The osmotic pressure of the internal electrolyte solution should be adjusted when the electrolyte concentration varies significantly on both sides of the membrane. This is performed by either addition of an inert electrolyte such as potassium sulfate or potassium chloride to the internal electrolyte solution of the probe, or by dilution of the test solution.

(v) Known addition, calibration and Gran's plot techniques are applied. For high precision, the temperature of the sample and standard should be within ±5°C and efficient stirring is applied to avoid adherence of solid samples to the probe membrane.

Applications of the gas-sensing membrane probes to the determination of many gases as well as ionic species after conversion to the gaseous form have been reviewed (Table 2.8). Conversely, while it is possible to determine ionic species such as nitrite, sulfite and bicarbonate by using the gas-sensing membrane probes after conversion to the gas form, gaseous species can be determined by the ion-selective membranes after conversion to the ionic species. For instance, sulfur dioxide is absorbed in hydrogen peroxide solution and the sulfate ion is determined using the solid state lead ion-selective membrane electrode [386]. Nitrogen oxides are oxidized to the nitrate ion by hydrogen peroxide [387, 388] or ozone [389, 390], followed

Table 2.8. Some commercially available gas-sensing membrane probes and their applications

Probe	Manufacturer	Model No.	Application Gas analysis in:	Reference
Ammonia	E.I.L	8002-2	Feces	407
	Orion	95-10	Sewage	408
			Wines	409
			Beer	410
			Soils	411–413
			Fish	414
			Neutral water	411, 415, 416
			Sea water	415, 417, 418
			Waste water	419–421
			Boiler water	422–426
			Fresh water	392
			Air	427, 428
			Urea	429
			Biological fluids	430
			Plasma	431–434
			Organic compounds	435–438
			Urine	415
Carbon dioxide	Radiometer	E 5036	Blood	439
	E.I.L	19210	Tissue and bio-	446
	Orion	95-02	logical fluids	
	Instrumental		Clinical materials	441, 442
	Lab. Inc.		Boiler water	373
			α-amino acids	379
			Bacterial cultures	443
Hydrogen cyanide	Orion	95-06		
Hydrogen fluoride	Orion	95-09		
Hydrogen sulfide	Orion	95-16		
Nitrogen oxides	Orion	95-46	Soils	444
			Water	444
			Stack gas	445
			Fish	446
Sulfur dioxide	Orion	95-46	Petroleum Products	447
	E.I.L	8010-2	Wine	384, 448

by measurement of the ions produced using a coated wire nitrate electrode [387], or the liquid ion-exchanger membrane nitrate electrode [388, 389]. Carbon dioxide in serum is converted into sodium bicarbonate by absorption in a buffer at pH 8·4 and the bicarbonate ion formed is determined using a liquid membrane electrode prepared from 5% Aliquat 336 in

4′-butyl-2,2,12-trifluoroacetophenone [391]. Gas permeable membranes have been employed for measurements under non-equilibrium conditions [391a].

E. Interferences

Because of the gas permeable membranes of the gas probes, gaseous species that are able to penetrate through the membrane will affect the potential of the sensor. For example, oxidizing agents such as chlorine or bromine in the sample solution will pass through the membrane and destroy the internal electrolyte. Severe interference—when it occurs—renders the response of the probe to be sluggish, necessitating the renewal of the internal electrolyte in order to restore performance.

Ammonia probes suffer interferences only from volatile amines [392] and mercuric ions. However, the effect of the latter can be suppressed by complexation with iodide ions [393]. Sulfur dioxide probes suffer interferences from concentrated hydrochloric, hydrofluoric and acetic acids. With nitrogen oxide probes, carbon dioxide in concentrations above 10^{-3} M is the most important interferer affecting the sensitivity of the measurement. The carbon dioxide probe is usually operated at the pH range of 5–7; under these conditions acetic acid appears to be the only interfering species. However, this effect is small due to the strong association of acetic acid and water molecules [379]. In general, most of the gas-sensing membrane probes are free from interferences with proper choice of the type and concentration of the internal electrolyte and proper sample preparation.

F. Oxygen sensing membrane probes

The oxygen sensing membrane probes are based on a different principle owing to the fact that either the pressure of oxygen or the current of reduction to O^{2-} is measured. However, these probes resemble the above-mentioned gas sensing membrane probes in being composed of two electrodes and a gas permeable membrane. The most common types of oxygen probes consist of a silver anode in a saturated potassium chloride solution and a platinum cathode (Fig. 2.21). When the probe is immersed in the test solution and a potential is applied, the oxygen is electrolytically reduced:

$$\text{Anode: } 4\,Ag + 4\,Cl^- \rightarrow 4\,AgCl + 4\,e \qquad (52)$$

$$\text{Cathode: } 4\,H^+ + 4\,e + O_2 \rightarrow 2\,H_2O \qquad (53)$$

If the polarized voltage applied is in the range of 0·5–0·8 V, the current generated is proportional to the oxygen concentration in the medium.

Several types of oxygen probes have been described. A cell consisting of two enclosed chambers, one containing the sample and the other contain-

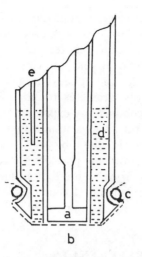

FIG. 2.21. Oxygen sensing membrane probe (The Clark type). a, platinum cathode; b, gas permeable membrane; c, rubber O-ring; d, saturated potassium chloride solution; e, silver anode.

ing an aqueous reference solution with a gas-permeable membrane between the two solutions, has been used [394]. The cathode is immersed in the reference solution and the anode is immersed in the sample test solution. A current is applied to the electrode to cause oxygen and hydrogen to liberate at the anode and cathode respectively, and the difference between pO_2 and pH_2 is measured by a differential pressure sensor coupled across the outlets of the two chambers. A thin-film pO_2 electrode has been constructed using gold paste [394a].

Membrane-type galvanic oxygen sensors [395, 396] have also been reported in which the oxygen diffuses through a semi-permeable plastic membrane (e.g. polypropene or PTFE) and is reduced electrochemically at a suitable cathode to O^{2-}. Various types of electrodes have been suggested. Electrodes made of porous silver–nickel alloy, such as those used in the fuel cells [397], or in the form of a tube of calcium oxide-activated zirconium dioxide closed at the end with an interior coating of platinum black [397], and cylindrical concentric electrodes made of zinc and silver amalgam [398], have been reported.

Oxygen sensors applicable at high temperatures, which are based on the use of nickel–nickel oxide, or platinum electrodes stabilized with antisintering materials such as zirconia, alumina, yttria and thoria [399, 400], as well as other oxides [401], have been described. Oxygen probes with fast response are prepared by mounting coils of lead wire anode and silver wire cathode on an Araldite rod using potassium bicarbonate as an electrolyte [402]. The system is sealed with a 25 μm polypropene membrane, and the output is fed to a digital voltammeter. Other oxygen probes are designed for biological study [403] or to collect large amounts of oxygen from the test solution [404]. An oxygen analyzer has been described to consist of a ZrO_2 ceramic tube closed at one end and coated both internally and externally with porous platinum to serve as an electrode. The tube is filled with air and the test gas circulates round its exterior. At any specific temperature, the ionic conductivity expressed in mV, due to displacement of O_2 in the lattice of ZrO_2, is logarithmically related to the ratio between pO_2 values inside and outside the tube [405].

Probes selective to hydrogen peroxide have been developed. These incorporate the oxygen probe with either a dialysis membrane containing lead oxide [406] or a regenerated cellulose [407] membrane containing Ru_2O_3 to decompose the hydrogen peroxide into oxygen gas.

REFERENCES

1. G. L. Vogel, L. C. Chow and W. E. Brown, *Anal. Chem.* **52**, 375 (1980).
2. M. Frant and J. Ross, *Science* **154**, 1553 (1966).
2a. A. Covington, *Chemistry in Britain* **5**, 389 (1969).
3. J. Lingane, *Anal. Chem.* **40**, 935 (1968).
4. A. Hulanicki, M. Trojanowicz and M. Cichy, *Talanta* **23**, 47 (1976).
5. British Patent No. 1258376 (10.2.1970).
6. U.S. Patent No. 3442782 (6.5.1969).
7. British Patent No. 1240028 (4.8.1969).
8. R. Durst and J. Taylor, *Anal. Chem.* **39**, 1483 (1967).
9. R. Durst, *Anal. Chem.* **41**, 2089 (1969).
10. M. Frant, U.S. Patent No. 3431182 (6.3.1969).
11. H. Bronstein and D. Manning, *J. Electrochem. Soc.* **119**, 125 (1972).
12. R. Robinson, W. Duer and R. Bates, *Anal Chem.* **43**, 1862 (1971).
13. R. Mesmer, *Anal. Chem.* **40**, 443 (1968).
14. J. Gatewood, Halliburton Services, Duncan, Oklahoma.
14a. J. Havas and L. Kecskes, *Magy. Kem. Foly.* **83**, 529 (1977).
15. T. Warner, *Anal. Chem.* **41**, 527 (1969).
16. J. Lingane, *Anal. Chem.* **39**, 881 (1967).
17. E. Baumann, *Anal. Chim. Acta* **42**, 127 (1968).
18. W. Hanson and D. Lloyd, *Chemy. Ind.* 1972, 41.
19. B. Ingram and I. May, Prof. Pap. U.S. Geol. Surv. 1971, No. 750-B and 180B.
20. J. Tusl, *J. Ass. Off. Anal. Chem.* **53**, 267 (1970).
21. J. Tusl, *Chemicke Listy* **64**, 322 (1970).
22. W. Selig, *Mikrochim. Acta* 1970, 229.
23. M. Frant, *Galvanotechnik* **63**, 745 (1972).
24. E. Baumann, *Anal. Chim. Acta* **54**, 189 (1971).
25. B. Ingram, *Anal. Chem.* **42**, 1825 (1970).
26. H. Zentner, *Chem. Ind.* 1973, 480.
27. R. Bock and S. Strecker, *Z. Anal. Chem.* **235**, 322 (1968).
28. T. Anfalt and D. Jagner, *Anal. Chim. Acta* **50**, 23 (1970).
29. P. Evans, G. Moody and J. Thomas, *Lab. Pract.* **20**, 644 (1971).

30. H. Birkeland and G. Rolla, *Archs. Oral Biol.* **17**, 455 (1972).
30a. J. Havas and L. Kecskes, *Magy. Kem. Foly.* **83**, 535 (1977).
31. H. Ademetzova and R. Vadura, *J. Electroanal. Chem.* **55**, 53 (1974).
32. D. Watson and D. Yee, *Electrochim. Acta* **14**, 1143 (1969).
33. D. Watson and D. Yee, *Electrochim. Acta* **16**, 549 (1971).
34. U.K. Patent No. 1150698 (30.4.1969).
35. J. Czaban and G. Rechnitz, *Anal. Chem.* **45**, 471 (1973).
36. Orion Research Inc., Instruction Manual (Halide electrodes) (1973).
37. N. Ogata, *Japan Analyst.* **21**, 780 (1972).
38. G. Lemahieu, C. Lemahieu-Hode and B. Rosibois, *Analusis* 1972, 110.
39. W. Ficklin and W. Gotschall, *Anal. Lett.* **6**, 217 (1973).
40. J. Ross and M. Frant, *The Mechanism of Interference and Solid State Ion Selective Electrodes.* Pittsburg Conference on Anal. Chem. and Applied Spectroscopy, March 1968.
41. E. Duff and J. Stuart, *Chem. Ind.* 1973, 1115.
42. R. Leest, *Analyst* **101**, 433 (1976).
43. L. Lechner and I. Van Sekerka, *J. Electroanal. Chem.* **57**, 317 (1974).
44. P. Tseng and W. Gutknecht, *Anal. Lett.* **9**, 795 (1976).
44a. Y. G. Vlasov, D. E. Hackleman and R. P. Buck, *Anal. Chem.* **51**, 1570 (1979).
45. S. S. M. Hassan. Unpublished work.
46. A. Varduca, D. Virtosu, V. Marculescu and C. Luca, *Revta Chim.* **27**, 527 (1976).
47. J. Ross, In *Proceedings of Symposium on Ion Selective Electrodes*, p. 84 (1969).
48. B. Fleet and H. VonStorp, *Anal. Chem.* **43**, 1575 (1971).
49. D. Evans, *Anal. Chem.* **44**, 875 (1972).
50. Orion Research Inc.; Instruction Manual (Cyanide Electrode, Model 94-06) (1972).
51. M. Mascini, *Anal. Chem.* **45**, 614 (1973).
52. B. Gyorgy, L. Andre, L. Stehli and E. Pungor, In *Proceeding of the International Measurement Confederation on Electrochemical Sensors.* Veszprem, Hungary, p. 111 (1968).
53. M. Wronski, *Analyst* **84**, 668 (1959).
54. I. Sekerka and J. Lechner, *Wat. Res.* **10**, 479 (1976).
55. E. Pungor and K. Toth, *Analyst* **95**, 625 (1970).
56. M. Frant, J. Ross and J. Riseman, *Anal. Chem.* **44**, 2227 (1972).
57. *Orion Research Newsletter* **6**, 1 (1974).
58. Orion Research Inc., Instruction Manual (Thiocyanate Electrode Model 94-58) (1971).
59. Orion Research Inc., Instruction Manual (Silver Electrode Model 94-16) (1970).
60. W. Blaedel and D. Dinwiddie, *Anal. Chem.* **46**, 873 (1974).
61. J. Meer, G. Boef and W. Linden, *Anal. Chim. Acta* **79**, 27 (1975).
62. J. Meer, G. Boef and W. Linden, *Anal. Chim. Acta* **79**, 261 (1975).
63. V. Olson, J. Carr, R. Hargens and R. Force, *Anal. Chem.* **48**, 1228 (1976).
64. M. El-Taras, E. Pungor and G. Nagy, *Anal. Chim. Acta* **82**, 285 (1976).
65. D. Crombie, G. Moody and J. Thomas, *Talanta* **21**, 1094 (1974).
66. D. Midgley, *Anal. Chim. Acta* **87**, 19 (1976).
67. G. Oglesby, W. Duer and F. Millero, *Anal. Chem.* **49**, 877 (1977).
68. Orion Research Inc., Instruction Manual (Copper electrode Model 94-29) (1973).
69. Y. Fung and K. Fung, *Anal. Chem.* **49**, 497 (1977).
70. G. Johansson and K. Edstroem, *Talanta* **19**, 1623 (1972).
71. M. Brand, J. Millitello and G. Rechnitz, *Anal. Lett.* **2**, 523 (1969).
72. G. Rechnitz and N. Kenny, *Anal. Lett.* **2**, 395 (1969).
73. J. Meer, G. Boef and W. Linden, *Anal Chim. Acta* **85**, 317 (1976).

74. W. Blaedel and D. Dinwiddie, *Anal. Chem.* **47**, 1070 (1975).
75. N. Savvin, V. Shterman, A. Gordievskii and A. Syrchenkov, *Zav. Lab.* **37**, 1025 (1971).
76. H. Hirata and M. Inoue, British Patent No. 1249288 (16.4.1970).
77. T. Anfalt and D. Jagner, *Anal. Chim. Acta* **56**, 477 (1971).
78. E. Hansen and J. Ruzicka, *Anal. Chim. Acta* **72**, 365 (1972).
79. J. Ruzicka and E. Hansen, *Anal. Chim. Acta* **63**, 115 (1973).
80. Orion Research Inc., Instruction Manual (Lead Electrode Model 94-82) (1972).
81. Orion Research Inc., Instruction Manual (Cadmium Electrode Model 94-48A) (1968).
82. G. Rechnitz and N. Kenny, *Anal. Lett.* **3**, 259 (1970).
83. P. Kivalo, R. Virtanen, K. Wickstroem, M. Wilson, E. Pungor, K. Toth and G. Sundhdm, *Anal. Chim. Acta* **87**, 387 (1976).
84. P. Kivalo, R. Virtanen, K. Wickstroem, M. Wilson, E. Pungor, G. Horvai and K. Toth, *Anal. Chim. Acta* **87**, 401 (1976).
85. *Orion Research Newsletter* **2**, 41 (1970).
86. I. Popescu, E. Hopirtean, L. Savici and R. Vlad, *Revue roum. Chim.* **20**, 993 (1975).
87. K. Hiiro, T. Tanaka, A. Kawahara and Y. Kono, *Japan Analyst* **22**, 1072 (1973).
88. D. Weiss, *Z. Anal. Chem.* **262**, 28 (1972).
89. R. Leest, *Analyst* **102**, 509 (1977).
90. D. Müller, P. West and R. Müller, *Anal. Chem.* **41**, 2038 (1969).
91. A. Allam, G. Pitts and J. Hollis, *Soil Sci.* **114**, 456 (1972).
92. H. Clysters, F. Adams and F. Verbeek, *Anal. Chim. Acta* **83**, 27 (1976).
93. P. Tseng and W. Gutknecht, *Anal. Chem.* **48**, 1996 (1976).
94. G. Rechnitz, G. Fricke and M. Mohan, *Anal. Chem.* **44**, 1098 (1972).
95. E. Barry, R. Butler and V. Larrakas, *Radiochem. Radioanalyt. Lett.* **21**, 105 (1975).
96. I. Sekerka and J. Lechner, *Anal. Lett.* **9**, 1099 (1976).
97. H. Hirata and K. Higashiyama, *Talanta* **19**, 391 (1972).
98. British Patent No. 1310084 (2.3.1972).
99. H. Hiroshi and K. Higashiyama, *Anal. Chim. Acta* **57**, 476 (1971).
100. British Patent No. 1310085 (2.3.1972).
101. British Patent No. 1330443 (2.3.1972).
102. British Patent No. 1313989 (2.3.1972).
103. British Patent No. 1310089 (2.3.1972).
104. British Patent No. 1310087 (2.3.1972).
105. J. Vesely, *Colln Czech. Chem. Commun.* **36**, 3364 (1971).
106. British Patent No. 1310086 (2.3.1972).
107. A. Zhukov, A. Vishnyakov, Y. Urusov and A. Gordievskii, *Zh. Analit. Khim.* **30**, 1614 (1975).
108. British Patent No. 1310088 (2.3.1972).
109. British Patent No. 1280187 (13.8.1969).
110. C. Baker and I. Trachtenberg, *J. Electrochem. Soc.* **118**, 571 (1971).
111. R. Jasinski, I. Trachtenberg and G. Rice, *J. Electrochem. Soc.* **121**, 363 (1974).
112. R. Jasinski, G. Barna and I. Trachtenberg, *J. Electrochem. Soc.* **121**, 1575 (1974).
113. H. Hirata and K. Higashiyama, *Anal. Chim. Acta* **54**, 415 (1971).
114. H. Hirata and K. Higashiyama, *Z. Anal. Chem.* **257**, 104 (1971).
115. C. Liteanu, I. Popescu and V. Ciovirnache, *Talanta* **19**, 985 (1972).
116. H. Hirata and M. Ayuzawa, *Chemy Lett.* **1974**, 1451.
117. N. Bausova, W. Bamburov, L. Manakova and A. Sivoplyas, *Zh. Analit. Khim.* **28**, 2042 (1973).
118. M. Sharp and G. Johansson, *Anal. Chim. Acta* **54**, 13 (1971).

119. M. Sharp, *Anal. Chim. Acta.* **85**, 17 (1976).
120. M. Sharp, *Anal. Chim. Acta* **59**, 137 (1972).
121. M. Sharp, *Anal. Chim. Acta* **62**, 385 (1972).
122. M. Sharp, *Anal. Chim. Acta* **61**, 99 (1972).
123. T. Nomura and G. Nakagawa, *Anal. Lett.* **8**, 873 (1975).
124. J. Havas, E. Papp and E. Pungor, *Magy. Kem. Foly.* **73**, 292 (1967).
125. J. Havas, E. Papp and E. Pungor, *Acta Chim. Acad. Sci. Hung.* **58**, 9 (1968).
126. A. Macdonald and K. Toth, *Anal. Chim. Acta* **41**, 99 (1968).
127. E. Buchanan and J. Seago, *Anal. Chem.* **40**, 517 (1968).
127a. G. Papeschi, S. Bordi and M. Carla, *J. Electrochem. Soc.* **125**, 1807 (1978).
128. E. Pungor, ACEMA Symposium (1964).
129. E. Pungor, *Anal Chem.* **39**, 29A (1967).
130. British Patent No. 1285655 (16.2.1970).
131. A Covington, *In* R. Durst (ed.), *Ion-Selective Electrodes.* National Bureau of Standards Special Publication, Washington, p. 89 (1969).
132. E. Pungor, J. Havas and K. Toth, *Anal. Chim. Hung.* **41**, 239 (1964).
133. E. Pungor, K. Toth and J. Havas, *Hung. Sci. Instr.* **3**, 2 (1965).
134. G. Rechnitz, M. Kresz and S. Zamochnick, *Anal. Chem.* **38**, 973 (1966).
135. G. Rechnitz and M. Kresz, *Anal. Chem.* **38**, 1787 (1966).
136. K. Toth and E. Pungor, *Anal. Chim. Acta* **51**, 221 (1970).
137. A. Macdonald, Reported at the International Symposium on Analytical Chemistry, Birmingham (July 1969).
138. G. Rechnitz, Z. Lin and S. Zamochnick, *Anal. Lett.* **1**, 29 (1967).
139. U. Lukkari, *Acta Chem. Fenn.* **45**, 182 (1972).
140. G. Guilbault and P. Brignac, *Anal. Chem.* **41**, 1136 (1969).
141. E. Pungor, E. Schmidt and K. Toth, Proc. IMEKO Symposium on Electrochemical Sensors, Veszprem, Hungary, p. 121 (1968).
142. H. Hirata and K. Date, *Bull. Chem. Soc. Japan* **46**, 1468 (1973).
143. J. Pick, K. Toth and E. Pungor, *Anal. Chim. Acta* **61**, 169 (1972).
144. J. Pick, K. Toth and E. Pungor, *Anal. Chim. Acta* **65**, 240 (1973).
145. H. Hirata and K. Date, *Talanta* **17**, 883 (1970).
146. H. Hirata and K. Date, *Anal. Chem.* **43**, 297 (1971).
147. E. Pungor and J. Havas, *Acta Chim. Hung.* **50**, 77 (1966).
148. J. Havas and E. Pungor, Proc. IMEKO Symposium on Electrochemical Sensors Veszprem, Hungary, p. 79 (1968).
149. C. Coetzee and A. Basson, *Anal. Chim. Acta* **57**, 478 (1971).
150. J. Pick, K. Toth, E. Pungor, M. Vasak and W. Simon, *Anal. Chim. Acta* **64**, 477 (1973).
151. A. Fogg, A. Pathan and T. Burns, *Anal. Lett.* **7**, 539 (1974).
152. P. Pock, T. Eyrich and S. Styer, *J. Electrochem. Soc.* **124**, 530 (1977).
153. N. Kazaryan and E. Pungor, *Magy. Kem. Foly.* **77**, 186 (1971).
154. N. Kazaryan and E. Pungor, *Anal. Chim. Acta* **60**, 193 (1972).
155. E. Hopirtean, C. Liteanu and A. Masa, *Revue roum. Chim.* **19**, 921 (1974).
156. G. Moody and J. Thomas, *Lab. Pract.* **23**, 475 (1974).
157. G. Moody and J. Thomas, *Chemy Ind.* 1974, 644.
158. J. Kratochvil, D. Band and T. Treasure, *Lab. Pract.* **26**, 102 (1977).
159. W. Morf, P. Wuhrmann and W. Simon, *Anal. Chem.* **48**, 1031 (1976).
160. G. Griffiths, G. Moody and J. Thomas, *Analyst*, **97**, 420 (1972).
161. R. Bloch, A. Shatkay and H. Saroff, *Biophys. J.* **7**, 865 (1967).
61a. L. Keil, G. J. Moody and J. D. R. Thomas, *Anal. Chim. Acta* **96**, 171 (1978).
162. G. Moody, R. Oke and J. Thomas, *Analyst* **95**, 910 (1970).

163. A. Shatkay, *Anal. Chem.* **39**, 1056 (1969).
164. R. Bloch, A. Shatkay and H. Saroff, *Biophys. J.* **7**, 865 (1967).
165. A. Craggs, L. Keil, G. Moody and J. Thomas, *Talanta* **22**, 10 (1975).
166. D. Ammann, M. Gueggi, E. Pretsch and W. Simon, *Anal. Lett.* **8**, 709 (1975).
167. G. Christoffersen and E. Johansen, *Anal. Chim. Acta* **81**, 191 (1976).
168. M. Brown, J. Pemberton and J. Owen, *Anal. Chim. Acta* **85**, 261 (1976).
168a. A. M. Y. Jaber, G. J. Moody and J. D. R. Thomas, *Analyst* **102**, 943 (1977).
169. H. James, Rep. U.K. Atom. Energy Auth., AEEW-R795 (1972).
170. T. Nomura and G. Nakagaw, *Japan Analyst* **20**, 1570 (1971).
171. S. Lal and G. Christian, *Anal. Chem.* **43**, 410 (1971).
172. W. Szczepaniak, J. Malicka and K. Ren, *Chemia analit.* **20**, 1141 (1975).
173. D. Ryan and M. Cheung, *Anal. Chim. Acta* **82**, 409 (1976).
174. J. Davies, G. Moody, W. Price and J. Thomas, *Lab. Pract.* **22**, 20 (1973).
175. G. Baum and M. Lynn, *Anal. Chim. Acta* **65**, 393 (1973).
176. M. Semler and H. Adametzova, *J. Electroanalyt. Chem.* **49**, 155 (1974).
177. B. Nikol'skii, E. Materova, A. Grekovich and V. Yurinskaya, *Zh. Analit. Khim.* **29**, 205 (1974).
178. I. Stepanova and R. Vadusa, *Chemicke Listy* **68**, 853 (1974).
179. O. Ryba and J. Petranek, *J. Electroanalyt. Chem.* **44**, 425 (1973).
180. J. Petranek and O. Ryba, *Anal. Chim. Acta* **72**, 375 (1974).
181. O. Ryba and J. Petranek, *Talanta* **23**, 158 (1976).
182. M. Gueggi, U. Fiedler, E. Pretsch and W. Simon, *Anal Lett.* **8**, 857 (1975).
183. E. Materova, A. Grekovich and N. Garbuzova, *Zh. Analit. Khim.* **29**, 1900 (1974).
184. J. Davies, G. Moody and J. Thomas, *Analyst* **97**, 87 (1972).
185. A. Hulanicki and R. Lewandowski, *Chemia analit.* **19**, 53 (1974).
186. T. Rohm and G. Guilbault, *Anal. Chem.* **46**, 590 (1974).
187. K. Hiiro, G. Moody and J. Thomas, *Talanta* **22**, 10 (1975).
188. M. Cakrt, J. Bercik and Z. Hladky, *Z. Anal. Chem.* **281**, 295 (1976).
189. O. Takaishvili, E. Motsonelidze, Yu. Karachentseva and P. Davitaya, *Zh. analit. Khim.* **30**, 1629 (1975).
190. T. Higuchi, C. Illian and J. Tossounian, *Anal. Chem.* **42**, 1674 (1970).
191. T. Tanaka, K. Hiiro and A. Kawahara, *Anal. Lett.* **7**, 173 (1974).
192. H. James, G. Carmack and H. Freiser, *Anal. Chem.* **44**, 856 (1972).
193. Y. Shijo, *Bull. Chem. Soc. Japan* **48**, 1647 (1975).
194. R. Cattrall, S. Tribuzio and H. Freiser, *Anal. Chem.* **46**, 2223 (1974).
195. R. Cattrall, D. Drew and I. Hamilton, *Anal. Chim. Acta* **76**, 269 (1975).
196. A. Ansaldi and S. Epstein, *Anal. Chem.* **45**, 595 (1973).
197. E. Hopirtean and E. Stefaniga, *Revue roum. Chim.* **20**, 863 (1975).
198. R. Caattrall and C. Pui, *Anal. Chem.* **47**, 93 (1975).
199. R. Cattrall and C. Pui, *Anal. Chim. Acta* **87**, 419 (1976).
200. R. Cattrall and C. Pui, *Anal. Chim. Acta* **88**, 185 (1977).
201. R. Cattrall and C. Pui, *Anal. Chem.* **48**, 552 (1976).
202. R. Cattrall and C. Pui, *Anal. Chim. Acta* **83**, 355 (1976).
203. S. Mesaric and E. Dehmen, *Anal. Chim. Acta* **64**, 431 (1973).
204. G. Qureshi and J. Lindquist, *Anal. Chim. Acta* **67**, 243 (1973).
205. J. Sapio, J. Colaruotolo and J. Bobbitt, *Anal. Chim. Acta* **67**, 240 (1973); **71**, 222 (1974).
206. A. Burdin, J. Mesplede and M. Porthault, *C. R. hebd. Seanc. Acad. Sci.*, Paris, **C276**, 65 (1973).
207. J. Ruzica, C. Lamm and J. Tjell, Ger. Offen. No. 2034686 (11.2.1971).
208. G. Baiulescu and V. Cosofret, *Revta Chim.* **26**, 1051 (1975).

9. V. Cosofret, *Revta Chim.* **27**, 240 (1976).
10. J. Ruzicka and C. Lamm, *Anal. Chim. Acta* **54**, 1 (1971).
11. A. Rouchouse, J. Mesplede and M. Porthoult, *Anal. Chim. Acta* **74**, 155 (1975).
12. J. Ruzicka and J. Tjell, *Anal. Chim. Acta* **49**, 346 (1970).
13. C. Luca, G. Semenescu and C. Nedea, *Revta Chim.* **25**, 1015 (1974).
14. G. Popa, G. Birsen and C. Luca, *Revta Chim.* **26**, 512 (1975).
15. E. Neiman, Y. Figel'son and V. Yakovleva, *Zh. analit. Khim.* **30**, 1977 (1975).
16. E. Pungor and E. Szepesvary, *Hung. Sci. Instr.* **15**, 9 (1968).
17. E. Pungor and E. Szepesvary, *Anal. Chim. Acta* **43**, 289 (1968).
18. E. Pungor, E. Szepesvary and J. Havas, *Anal. Lett.* **1**, 213 (1968).
19. G. Farsang, *Hung. Sci. Instr.* **12**, 22 (1968).
20. A. Weser and E. Pungor, Proc. IMEKO Symposium on Electrochemical Sensors, Veszprem, Hungary, p. 99 (1968).
21. E. Pungor, Z. Feher and G. Nagy, *Magy. Kem. Foly* **77**, 289 (1971).
22. E. Pungor, G. Nagy and Z. Feher, *Magy. Kem. Foly* **77**, 294 (1971).
23. A. Weser and E. Pungor, *Acta Chim.* (*Budapest*) **64**, 311 (1970).
24. O. Kabanova and Y. Goncharov, *Zh. Analit. Khim.* **28**, 1665 (1973).
25. J. Ruzicka, C. Lamm and J. Tjell, *Anal. Chim. Acta* **62**, 15 (1972).
26. M. Mascini, F. Pallozzi and A. Liberti, *Anal. Chim. Acta* **64**, 126 (1973).
27. E. Szepesvary and E. Pungor, *Anal Chim. Acta* **54**, 199 (1971).
28. E. Szepesvary and E. Pungor, *Anal. Lett.* **3**, 603 (1970).
29. E. Pungor and E. Szepesvary, *Periodica polytech. Chem. Eng.* **16**, 323 (1972).
30. M. Mascini and A. Liberti, *Anal. Chim. Acta* **51**, 231 (1970).
31. M. Mascini and A. Liberti, *Anal. Chim. Acta* **64**, 63 (1973).
32. M. Mascini and A. Liberti, *Anal. Chim. Acta* **53**, 202 (1971).
33. J. VanLoon, *Anal. Chim. Acta* **54**, 23 (1971).
34. E. Materova and S. Mikhailova, *Vestn. Leningrad Univ., Fiz., Khim.* 1971, 43.
4a. G. J. M. Heijne, W. E van der Linden and G. den Boef, *Anal. Chim. Acta* **100**, 193 (1978).
35. A. Grekovich, E. Materova and N. Garbuzova, *Zh. analit. Khim.* **28**, 1206 (1973).
36. J. Parsons, *Anal. Chem.* **30**, 1262 (1958).
37. W. Malik, S.Srivastava, V. Bhandari and S. Kumar, *J. Colloid Interface Sci.* **47**, 1 (1974).
38. M. Adhikari, D. Gangopadhyay and G. Sen, *J. Indian. Chem. Soc.* **52**, 1136 (1975).
39. F. Shu and G. Guilbault, *Anal. Lett.* **5**, 559 (1972).
40. A. Davies, Rep. U.K. atom. Energy Auth. AWRE-O 46/71 (1971).
41. M. Mascini and A. Liberti, *Anal. Chim. Acta* **47**, 339 (1969).
42. M. Mascini, *Anal. Chim. Acta* **62**, 29 (1972).
43. O. Schaefer, *Anal. Chim. Acta* **87**, 495 (1973).
44. C. Coetzee and A. Basson, *Anal. Chim. Acta* **64**, 300 (1976).
45. W. Malik, S. Srivastava, P. Razdan and S. Kumar, *J. Electroanalyt. chem.* **72**, 111 (1976).
46. S. Lal, *Z. Anal. Chem.* **255**, 209 (1971).
47. S. Lal, *Z. Anal. Chem.* **255**, 210 (1971).
48. S. Lal and G. Christian, *Naturwissenschaften* **58**, 362 (1971).
49. S. Lal and G. Christian, *Anal. Chem.* **43**, 410 (1971).
50. E. Materova, G. Grinberg and M. Evstifeeve, *Zh. Analit. Khim.* **24**, 821 (1969).
51. O. LeBlanc and W. Grubb, *Anal. Chem.* **48**, 1658 (1976).
52. H. Tendeloo and A. Krips, *Rec. Trav. Chim.* **76**, 703 (1957).
53. H. Tendeloo and A. Krips, *Rec. Trav. Chim.* **76**, 946 (1957).
54. H. Tendeloo and A. Krips, *Rec. Trav. Chim.* **78**, 177 (1959).
55. H. Tendeloo and Van derVoort, *Rec. Trav. Chim.* **79**, 639 (1960).

256. A. Rajput, M. Kataoka and T. Kambara, *J. Electroanalyt. Chem.* **66**, 67 (1975).
257. P. Cloos and J. Fripiat, *Bull. Soc. Chim. Fr.* 1960, 423.
258. A. Shatkay, *Anal. Chem.* **39**, 1056 (1967).
259. R. Bloch, A. Shatkay and H. Saroff, *Biophys. J.* **7**, 865 (1967).
260. E. Buchanan and J. Seago, *Anal. Chem.* **40**, 517 (1968).
261. A. Covington and N. Kumar, *Anal. Chim. Acta* **85**, 175 (1976).
262. S. Lewis and R. Buck, *Anal. Lett.* **9**, 439 (1976).
263. R. Fischer and R. Babcock, *Anal. Chem.* **30**, 1732 (1958).
264. P. Hirsch-Ayalon, *J. Polym. Sci.* **23**, 697 (1957); *Rec. Trav. Chim.* **79**, 382 (1960).
265. C. Liteanu and M. Mioscu, *Talanta* **18**, 51 (1971).
266. C. Liteanu, M. Mioscu and I. Popescu, *Studia Univ. Babes. Bolyai, Ser. Chem.* **16**, 25 (1971).
267. C. Liteanu and I. Popescu, *Talanta* **19**, 974 (1972).
268. F. Siddiqi, N. Lakshminarayanaiah and S. Saxena, *Z. Phys. Chem.* **72**, 298 (1970).
269. E. Hopirtean, E. Stefaniga, C. Liteanu and I. Gusan, *Revta Chim.* **27**, 346 (1976).
270. E. Hopirtean, E. Stefaniga and C. Liteanu, *Chemia analit.* **21**, 867 (1976).
271. E. Hopirteau, M. Preda and C. Liteanu, *Chemia analit.* **21**, 861 (1976).
271a. H. A. Laitinen and T. M. Hseu, *Anal. Chem.* **51**, 1550 (1979).
272. N. Kamo, N. Hazemoto and Y. Kobatake, *Talanta* **24**, 111 (1977).
273. J. Ross, *Science* **156**, 1378 (1967).
274. W. Szczepaniak and K. Ren, *Anal. Chim. Acta* **82**, 37 (1976).
275. A. Fogg, A. Al-Sibaai and C. Burgess, *Anal. Lett.* **8**, 129 (1975).
276. A. Fogg and A. Al-Sibaai, *Anal. Lett.* **9**, 33 (1976).
277. Y. Shijo, *Bull. Chem. Soc. Japan* **48**, 1647 (1975).
278. M. Novkirishka and R. Christova, *Anal. Chim. Acta* **78**, 63 (1975).
279. F. Orme, in M. Lavallee, O. F. Schanne and N. C. Hebert (eds), *Glass Microelectrodes*. Wiley, New York, p. 388 (1968).
280. M. Oehme and W. Simon, *Anal. Chim. Acta* **86**, 21 (1976).
281. U.S. Patent No. 3467591 (16.9.1969).
282. G. Scibona, L. Mantella and P. Danesi, *Anal. Chem.* **42**, 844 (1970).
283. Belgian Patent No. 668409 (17.2.1966).
284. U.S. Patent No. 3406102 (15.10.1968).
285. U.S. Patent No. 3467590 (16.9.1969).
286. D. Jagner and J. Oestergaard-Jensen, *Anal. Chim. Acta* **80**, 9 (1975).
286a. G. J. Moody, N. S. Nassory and J. D. R. Thomas, *Analyst* **103**, 68 (1978).
287. R. Huston and J. Butler, *Anal. Chem.* **41**, 200 (1969).
288. J. Bagg and R. Vinen, *Anal. Chem.* **44**, 1773 (1972).
289. J. Leyendekkers and M. Whitfield, *Anal. Chem.* **43**, 322 (1971).
290. M. Whitfield and J. Leyendekkars, *Anal. Chem.* **42**, 444 (1970).
291. R. Burr, *Clinica Chim. Acta* **43**, 311 (1973).
292. R. Geyer and I. Preuss, *Z. Chemie, Lpz.* **14**, 368 (1974).
293. British Patent No. 1284476 (29.5.1970).
294. C. Coetzee and A. Basson, *Anal. Chim. Acta* **83**, 361 (1976).
295. R. Levins, *Anal. Chem.* **43**, 1045 (1971).
296. E. Baumann, *Anal. Chem.* **47**, 959 (1975).
297. A. Gordievskii, A. Syrchenkov, S. Savvin, V. Shterman, S. Chizhevskii and A. Zhukov, *Trudy Mosk. Khim. Tekhnol Inst.* 1972, 140.
297a. T. Malone and G. Christian, *Anal. Lett.* **7**, 33 (1974).
298. E. Hopirtean and E. Stefaniga, *Chem. Analit. (Warsaw)* **22**, 845 (1977).
299. R. Carlson and J. Paul, *Anal. Chem.* **40**, 1292 (1968).
300. C. Gavach and J. Guastalla, *Membranes Permeabilite Selec., Colloq.* 1967, 165.

01. C. Gavach and C. Bertrand, *Anal. Chim. Acta* **55**, 385 (1971).
02. G. Scibona, P. Danesi, A. Conte and B. Scuppa, *J. Colloid Interface Sci.* **35**, 631 (1971).
03. M. Matsui and H. Freiser, *Anal. Lett.* **3**, 161 (1970).
04. J. Sandblom, G. Eisenman and J. Walker, *J. Phys. Chem.* **71**, 3862, 3871 (1967).
05. O. Bonner and D. Lunney, *J. Phys. Chem.* **70**, 1140 (1966).
06. G. Scibona, P. Danesi, F. Salvemini and B. Scuppa, *J. Phys. Chem.* **75**, 54 (1971).
07. G. Scibona and B. Scuppa, *Corsi Semin. Chim.* **1968**, 23.
08. C. Fabiani, *Anal. Chem.* **48**, 865 (1976).
09. A. Gordievskii, A. Syrchenkov, N. Savvin, V. Shterman and G. Kozhukhova, *Zav. Lab.* **38**, 265 (1972).
10. P. Danesi, G. Scibona and B. Scuppa, *Anal Chem.* **43**, 1892 (1971).
11. A. Grekovich, E. Materova and N. Garbuzova, *Zh. Analit. Khim.* **28**, 1206 (1973).
12. H. Herman and G. Rechnitz, *Science, N.Y.* **184**, 1070 (1974).
12a. L. A. Shumilova, A. V. Gordievskii, Y. A. Klyachko and N. G. Sarishvili, *Zh. Anal. Khim.* **32**, 2368 (1977).
13. A. Grekovich, E. Materova and V. Yurinskaya, *Zh. Anal. Khim.* **27**, 1218 (1972).
14. T. Shigematsu, A. Ota and M. Matsui, *Bull. Inst. Chem. Res. Kyoto Univ.* **51**, 268 (1973).
15. C. Coetzee and H. Freiser, *Anal. Chem.* **40**, 2071 (1968).
16. S. Back, *Anal. Chem.* **44**, 1696 (1972).
17. J. Vosta and J. Havel, *Scripta*, 1973, 3 *Chemia* (3), 109.
17a. S. K. Norov, V. V. Palchevskii and E. S. Gureev, *Zh. Anal. Khim.* **32**, 2394 (1977).
18. W. Haynes and J. Wagenknecht, *Anal. Lett.* **4**, 491 (1971).
18a. T. P. Hadjiioannou and E. P. Diamandis, *Anal. Chim. Acta* **94**, 443 (1977).
19. N. Ishibashi, H. Kohara and K. Horinouchi, *Talanta* **20**, 867 (1973).
20. A. Fogg, A. Pathan and D. Burns, *Anal. Chim. Acta* **73**, 220 (1974).
20a. H. Kohara, *Kitakyushu Kogyo Koto Semmon Gakko Kenkyn Hokoku* **11**, 159 (1978).
20b. E. Hopirtean and E. Stefaniga, *Rev. Roum. Chim.* **23**, 137 (1978).
21. A. Fogg, J. Kumar and D. Burns, *Anal. Lett.* **7**, 629 (1974).
22. E. Skobets, L. Makovestskaya and Y. Makovetskii, *Zh. Analit. Khim.* **29**, 2354 (1974).
23. A. Syrchenkov, Y. Urusov, M. Geminova, A. Zhukov, N. Savvin and A. Gordievskii, *Zav. Lab.* **40**, 1041 (1974).
24. M. Nanjo, T. Rohm and G. Guilbault, *Anal. Chim. Acta* **77**, 19 (1975).
25. A. Wilson and K. Pool, *Talanta* **23**, 387 (1976).
26. A. Bychkov, O. Petrukhin, V. Zarinskii, Yu. Zolotov, L. Bakhtinova and G. Shanina, *Zh. Analit. Khim.* **31**, 2114 (1976).
27. A. Bychkov, O. Petrukhin, V. Zarinskii and Yu. Zolotov, *Zh. Analit. Khim.* **30**, 2213 (1975).
28. R. Reinsfelder and F. Schultz, *Anal. Chim. Acta* **65**, 433 (1973).
29. J. Ross, U.S. Patent No. 3483112 (9.12.1969).
30. N. Ishibashi and H. Kohara, *Anal. Lett.* **4**, 785 (1971).
31. J. Ross, U.S. Patent No. 3497424 (24.2.1970).
32. E. Gibson, S. Shiller, J. Riseman, U.S. Patent No. 3467590 (16.9.1969).
33. J. Ross and M. Frant, *Liquid Ion Exchange Membrane Electrodes*. Pittsburg Conference on Analytical Chem. and Applied Spectroscopy, March 1966.
34. E. Materova, V. Muchovikov and M. Grigor'eva, *Anal. Lett.* **8**, 167 (1975).
35. J. Harrell, A. Jones and G. Choppin, *Anal. Chem.* **41**, 1459 (1969).
35a. C. R. Martin and H. Freiser, *Anal. Chem.* **52**, 562 (1980).

336. B. Birch and D. Clarke, *Anal. Chim. Acta* **61**, 159 (1972).
337. M. Sharp, *Anal. Chim. Acta* **65**, 405 (1973).
338. A. Gordievskii, V. Shterman, A. Syrchenkov, N. Savvin and A. Zhukov, *Zh. Analit. Khim.* **27**, 772 (1972).
339. N. Evseeva, I. Kremenskaya, V. Golubev and S. Timofeeva, *Zh. Analit. Khim.* **31**, 822 (1976).
340. M. Kataoka and T. Kambara, *J. Electroanalyt. Chem.* **73**, 279 (1976).
341. Z. Stefanac and W. Simon, *Microchem. J.* **12**, 125 (1967).
342. W. Simon, H. Wuhrmann, M. Vasak, L. Pioda, R. Dohner and Z. Stefanac, *Angew. Chem.* **82**, 433 (1970); *Angew. Chem. Internat. Ed.* **9**, 445 (1970).
343. L. Pioda, V. Stankova and W. Simon, *Anal. Lett.* **2**, 665 (1969).
344. M. Frant and J. Ross, *Science, N.Y.* **167**, 987 (1970).
345. S. Lal and G. Christian, *Anal. Lett.* **3**, 11 (1970).
346. R. Scholer and W. Simon, *Chimica* **24**, 372 (1970).
347. W. Simon, Swiss Patent No. 479870 (28.11.1969).
348. R. Cosgrove, C. Mask and I. Krull, *Anal. Lett.* **3**, 457 (1970).
349. I. Krull, C. Mask and R. Cosgrove, *Anal. Lett.* **3**, 43 (1970).
350. British Patent No. 1289916 (8.6.1971).
351. L. Vorob'ev and Y. Khitrov, *Stud., Biophys.* **26**, 49 (1971).
352. I. Krull, C. Mask and R. Cosgrove, *Anal. Lett.* **3**, 43 (1970).
353. E. Eyal and G. Rechnitz, *Anal. Chem.* **43**, 1090 (1971).
354. H. Seto, A. Jyo and N. Ishibashi, *Chemy Lett.* 1975, 483.
355. S. Hammond and P. Lambert, *J. Electroanal. Chem.* **53**, 155 (1974).
356. H. Wasserman, J. Mckeon, L. Smith and P. Forgione, *Tetrahedron Suppl. No.* **8**, Part II, 647 (1966).
357. K. Harashima, N. Tsuchida, T. Tanaka and J. Nagatsu, *Agr. Biol. Chem.* **31**, 481 (1967).
358. O. Kedem, E. Loebel and M. Furmansky, Ger. Offen. No. 2027128 (23.12.1970).
359. U. Fiedler, *Anal. Chim. Acta* **89**, 111 (1977).
360. J. Dye, M. Lok, F. Tehan, R. Coolen, N. Papadakis, J. Ceraso and M. DeBacker, *Ber. Bunsenges Phys. Chem.* **75**, 659 (1971).
361. H. Frensdorff, *J. Amer. Chem. Soc.* **93**, 600 (1971).
362. J. Lehn and J. Sauvage, *J. Chem. Soc.* (D) 1971, 440.
363. H. Frensdorff, *J. Amer. Chem. Soc.* **93**, 4684 (1971).
364. G. Rechnitz and E. Eyal, *Anal. Chem.* **44**, 370 (1972).
365. R. Levins, *Anal. Chem.* **43**, 1045 (1971).
366. W. Hildebrandt and K. Pool, *Talanta* **23**, 469 (1976).
367. J. Ross, J. Riseman and J. Krueger, *Pure Appl. Chem.* **36**, 473 (1973).
368. R. Stow, R. Baer and B. Randall, *Arch. Phys. Med. Rehabil.* **38**, 646 (1957).
369. J. Severinghaus and A. Bradley, *J. Appl. Physiol.* **13**, 515 (1958).
370. K. Gertz and H. Loeschcke, *Naturwissenschaften* **45**, 160 (1958).
371. M. Okada and H. Matsusita, *J. Chem. Soc. Japan Ind. Chem. Sect.* **72**, 1407 (1969).
372. *Orion Newsletter* **5**, 7 (1973).
373. D. Midgley, *Analyst* **100**, 386 (1975).
374. P. Bailey and M. Riley, *Analyst* **100**, 145 (1975).
375. P. Bailey and M. Riley, *Analyst* **102**, 213 (1977).
376. E. Hanson and N. Larsen, *Anal. Chim. Acta* **78**, 459 (1975).
377. K. Selinger, R. Staroscik and F. Malecki, *Chemia analit.* **21**, 1153 (1976).
378. T. Anfalt, A. Graneli and D. Jagner, *Anal. Chim. Acta* **76**, 253 (1975).
379. G. Guilbault and F. Shu, *Anal. Chem.* **44**, 2161 (1972).
379a. M. A. Jensen and G. A. Rechnitz, *Anal. Chem.* **51**, 1972 (1979).

380. P. Todd, *J. Sci. Fd. Agric.* **24**, 488 (1973).
380a. G. Buckee, *J. Inst. Brew.* **80**, 291 (1974).
381. T. S. Ma and R. C. Rittner, *Modern Organic Elemental Analysis.* Dekker, New York, p. 88 (1979).
382. J. Ruzicka and E. Hansen, *Anal. Chim. Acta* **69**, 129 (1974).
383. *Orion Newsletter* **6**, 3 (1974).
384. A. Binder, S. Ebel, M. Kaal and T. Thron, *Dt. Lebensmitte Rdsch* **71**, 246 (1975).
385. J. Mertens, P. VanderWinkel and D. Massart, *Anal. Lett.* **6**, 81 (1973).
386. M. Young, J. Driscoll and K. Mahoney, *Anal. Chem.* **45**, 2283 (1973).
387. B. Kneebone and H. Freiser, *Anal. Chem.* **45**, 449 (1973).
388. R. DiMartini, *Anal. Chem.* **42**, 1102 (1970).
389. J. Driscoll, A. Berger, J. Becker, J. Funkhouser and J. Valentine, 64th Annual Meeting of Air Pollution Control Assn., June 1971.
390. Y. Katagiri, T. Shimida, S. Fukui and S. Kanno, *J. Hyg. Chem.* **20**, 322 (1974).
391. H. Herman and G. Rechnitz, *Anal. Lett.* **8**, 147 (1975).
391a. I. Sekerka and J. F. Lechner, *Anal. Chim. Acta* **93**, 129 (1977).
392. M. Beckett and A. Wilson, *Wat. Res.* **8**, 333 (1974).
393. British Patent No. 1360220 (24.9.1971).
394. British Patent No. 1391168 (21.4.1973).
394a. S. Makinoda, S. Miura, and T. Koyama, *Experienta* **34**, 62 (1978).
395. British Patent No. 1200595 (19.10.1966).
396. British Patent No. 1392173 (9.3.1971).
397. Y. Rybkin and A. Seredenko, *Ukr. khim. Zh.*, **40**, 489 (1974).
398. British Patent No. 1223363 (11.3.1969).
399. British Patent No. 1217625 (3.5.1969).
400. British Patent No. 1201806 (23.12.1967).
401. A. Tseung and H. Bevan, *J. Electroanalyt. Chem.* **45**, 429 (1973).
402. C. Hahn, *J. Appl. Physiol.* **37**, 439 (1974).
403. British Patent No. 1359975 (19.2.1973).
404. British Patent No. 1226610 (18.7.1968).
405. G. Challet, *Verres Refract.* **30**, 365 (1976).
406. K. Shick, V. Magearu, N. Field and C. Huber, *Anal. Chem.* **48**, 2186 (1976).
407. E. Byrne and T. Power, *Comm. Soil Sci. Plant Anal.* **5**, 51 (1974).
408. D. Rogers and K. Pool, *Anal. Lett.* **6**, 801 (1973).
409. D. McWilliam and C. Ough, *Amer. J. Enol. Viticult* **25**, 67 (1974).
410. F. Drawert and T. Nitsche, *Brouwissenschaft* **29**, 299 (1976).
411. W. Banwart, M. Tabatabai and J. Bremner, *Comm. In Soil. Sci. and Plant Anal.* **3**, 449 (1972).
412. E. Busenberg and C. Clemency, *Clays and Clay Minerals* **21**, 213 (1973).
413. G. Miller, F. Riecken and N. Walter, *Soil. Sci. Soc. Amer. Proc.* **39**, 372 (1975).
414. J. Barica, *J. Fisheries Res. Bd. Canada* **30**, 1389 (1973).
415. R. Thomas and R. Booth, *Environ. Sci. Technol.* **7**, 523 (1973).
416. L. Vandevenne and J. Oudewater, Centre Belge d'Elude Doc. Eaux, Report No. 352, p. 127 (1973).
417. T. Gilbert and S. Clay, *Anal. Chem.* **45**, 1757 (1973).
418. A. Merks, *Neth. J. Sea Res.* **9**, 371 (1975).
419. P. LeBlanc and J. Sliwinski, Paper presented at the CIC-CCIW symposium on water Quality Parameters, Burlington, Ontario, Canada, Nov. 19, 1973.
420. J. Montalvo, *Anal. Chim. Acta* **65**, 189 (1973).
421. P. LeBlanc and J. Sliwinski, *Amer. Lab.* **5**, 51 (1973).
422. G. Hoge, H. Hazenberg and C. Gips, *Clinica Chim. Acta* **55**, 273 (1974).

423. D. Midgley and K. Torrance, *Analyst* **97**, 626 (1972).
424. D. Midgley and K. Torrance, *Analyst* **98**, 217 (1973).
425. J. Mertens, P. Van de Winkel and D. Massart, *Bull. Soc. Chim. Belg.* **83**, 19 (1974).
426. N. Shibata, *Anal. Chim. Acta* **83**, 371 (1976).
427. M. Fagan and L. DuBois, *Anal. Chim. Acta* **70**, 157 (1974).
428. B. Schreiber and R. Frei, *Mikrochim Acta* 1975, 219.
429. T. Woodis and J. Cummings, *J. Assoc. Offic. Agr. Chemists* **56**, 373 (1973).
430. J. Renfro and Y. Patel, *J. Appli. Physiol.* **37**, 756 (1974).
431. R. Coleman, *Clin. Chem.* **18**, 867 (1972).
432. G. Sanders and W. Thornton, *Clinica Chim. Acta* **46**, 465 (1973).
433. N. Park and J. Fenton, *J. Clin. Path.* **26**, 802 (1973).
434. A. Attili, D. Autizi, L. Capocaccia, S. Costantini and F. Cotta-Ramusino, *Biochem. Med.* **14**, 109 (1975).
435. J. Bremner and M. Tabatabai, *Comm. Soil Sci. Plant Anal.* **3**, 159 (1972).
436. A. Deschreider and R. Meaux, *Analusis* **2**, 442 (1973).
437. K. Maring and R. Steinmann, *Chem. Rundschau* **27**, 699 (1973).
438. J. Bremner and M. Tabatabai, *Comm. Soil Sci. Plant Anal.* **3**, 159 (1972).
439. L. Kempen, H. Deurenberg and F. Kreuzer, *Resp. Physiol.* **14**, 366 (1972).
440. C. Caflisch and N. Carter, *Anal. Biochem.* **60**, 252 (1974).
441. J. Severinghaus, *Ann. N.Y. Acad. Sci.* **148**, 115 (1968).
442. A. Smith and C. Hahn, *Br. Anaesth.* **41**, 731 (1969).
443. J. Ladenson, M. Huebnar and J. Marr, *Anal. Biochem.* **36**, 56 (1975).
444. M. Tabatabai, *Comm. Soil Sci. Plant Anal.* **5**, 569 (1974).
445. J. Driscoll, A. Berger, J. Becker, J. Funkhouser and J. Valentine, Paper presented at the 6th Annual Meeting of the Air Pollution Control Assn., Atlantic City, N.J., June 1971.
446. S. Sherkin, *J. Assoc. Offic. Anal. Chemists* **59**, 971 (1976).
447. J. Krueger, *Anal. Chem.* **46**, 1338 (1974).
448. E. Hansen, H. Filho and J. Ruzicka, *Anal. Chim. Acta* **71**, 225 (1974).

3. MEASUREMENT TECHNIQUES USING ION-SELECTIVE ELECTRODES

I. INTRODUCTORY REMARKS

The potential behavior of the ion-selective electrodes in solutions containing ionic species for which the electrodes are selective is described by the Nernst equation. Thus, the ion-selective electrodes sense the activity of the ion, which can be converted to concentration by taking the activity coefficient of the sample solution into account. These electrodes are able to measure the activity with considerable sensitivity often to below one part in 10^9, and with high selectivity.

Analytical techniques used for concentration measurements with ion-selective electrodes may be classified into two categories: (a) direct potentiometric measurements based on the Nernstian logarithmic relationship between e.m.f. and concentration, and (b) potentiometric titration techniques. Measurements based on direct potentiometry (calibration, kinetic, known addition and known subtraction methods) are affected by the precision of the potential readings. A precision of ±1 mV can be attained with relatively little experience using an expanded scale pH-meter, but more precise readings within ±0·05 mV require extreme care and temperature control (±5°C ~ ±1 mV).

The accuracy of the results, in turn, depends mainly on the precision of the potential readings. The magnitude of error can be predicted from the differential form of the Nernst equation:

$$dE/S = dc/C \tag{1}$$

where dE is the precision of potential measurement, dc is the resulting precision in the determination of concentration, S is the Nernstian slope and C is the test ion concentration. A potential precision of ±1 mV will result in a precision in concentration for a monovalent ion ($S = 60$) amounting to ±1·7% ($1/60 \times 100$), and for a divalent ion ($S = 30$), the concentration precision is about 3·3% ($1/30 \times 100$).

On the other hand, in potentiometric titrations, the potential is measured as a function of volume of the titrant added. Precision in the potential readings of ± 10 mV usually gives accurate results within ± 0.1 to $\pm 0.3\%$ of the theoretical. This is due to the fact that the steep slope of the titration curves near the equivalence point is usually more than 100 mV, and the activity coefficient and liquid junction factors have no effect on the titration. Hence, the accuracy of the results depends only on the accuracy of volume measurements and standardization of the titrant.

In general, direct potentiometry is a great deal faster and is useful for measurements of an ion in the presence of many other interfering substances, while potentiometric titrations provide more accurate results at the expense of time.

In order to utilize the whole working range of ion-selective electrodes, it is often preferred to measure the potentials at various concentrations to get a straight line of potentials versus logarithm of concentration. Besides, the numerical evaluation of the electrode parameters generally yields more statistical errors of these parameters. Using error propagation, it is possible to calculate the error in analysis caused by errors in calculation and measurement [1].

II. CALIBRATION AND KINETIC METHODS

A. Calibration graphs and electrode calibration

Empirical calibration graphs, whereby the electrode potential is related to the activity or concentration of the ion of interest, are generally used. The potential is plotted on semi-log graph paper and a linear calibration graph is constructed for standard solutions; then the unknown is compared with the graph. The slope of the curve depends not only on the ionic charge of the ion to be determined but also on the electrode behavior [2–6]. Mono- and divalent ions show slopes of about 59 and 29.5 mV per concentration decade, respectively. However, with the iodide electrode the mercuric ions show a calibration graph with a slope of 59 mV although the ion to be sensed is divalent [7].

The calibration graphs are made from completely pure dissociated salts containing the ion of interest in concentrations covering the range to be studied. However, calibration graphs can be drawn to cover six decades of concentrations so that samples differing in concentration by decades can be measured one after another. This method is valid assuming that there are no interferences due to ion binding and that the electrode is unable to sense ions other than the relevant ions in the test solution, and furthermore that the activity coefficient is approximately in unity or identical in both the standard and test solutions. In the presence of indifferent electrolytes, errors are introduced due to the variation in the ionic strength and the activity coefficient. This effect is tolerated either by calculation or by adjusting the ionic strength of both the standard and the test solutions to be identical.

For simple systems where the interference ions are known and their activity coefficient data are available, correction for the ionic strength effects can be easily made in order to relate the observed potential and the activity to concentration. A nomograph based on such relations is available for divalent ions [8]. However, this technique can be misleading for solutions of high or low pH values, where the high ionic mobilities of the hydrogen or hydroxyl ions constitute a major contribution to the ionic strength.

For complex systems, it is necessary to use calibration solutions which either approximate the background of the test solution, or contain an excess of an inert electrolyte or buffer solution [9]. Addition of an indifferent electrolyte to both the test and standard solutions may also be used in order to bring them to the same ionic strength, so that the activity coefficient of the test ions in all the solutions are identical. It is necessary that the contribution to the ionic strength from the added indifferent electrolyte is greater compared to that present in the test solution. For this purpose, total ionic strength adjustor buffer solutions (TISAB's) are used whereby the ionic strength is fixed by the high level of ions contained in these buffers. In general, this method is useful for divalent ions because of the greater deviation of the activity coefficient from unity for divalent ions as compared to monovalent ions.

The time required for constructing calibration graphs is saved by calibrating the electrode with two standard solutions which bracket the unknown concentration, and assuming a linear response of the electrode [10–12].

Using the equation of a straight line, the concentration of the unknown ion is easily calculated by the guide of the two standard solutions [12]. For example, if the potential readings for 10 μg and 100 μg of a standard fluoride solution are 137·5 and 78·3 mV respectively, and the potential of an unknown sample solution is 91·7 mV, then:

Fluoride concentration

$$= \text{antilog} \frac{\text{mV reading of 10 } \mu\text{g standard} - \text{mV reading of the unknown}}{\text{mV reading of 10 } \mu\text{g standard} - \text{mV reading of 100 } \mu\text{g standard}} \quad (2)$$

$$\text{Fluoride concentration of the unknown} = \text{antilog} \frac{137·5 - 91·7}{137·5 - 78·3} = 59·38 \ \mu\text{g} \quad (3)$$

B. Kinetic methods

Kinetic procedures for analysis using the ion-selective electrodes have become an important approach for the assay of inhibitors, catalysts, activators, substrates and enzymes [13–17].

Enzymes exhibit specificity with respect to a particular reaction and are considered as biological catalysts. This specificity and ability to catalyse reactions of substrates at low concentrations is of great use in the chemical analysis of both substrates and enzymes. In most of the enzymatic reactions ionic species are directly or indirectly produced [18–20].

The methods used for the determination of enzymes by ion-selective electrodes are based on the measurement of the rate of substrate consumption by following the concentration of the ionic species. The change in potential with time is recorded and the enzyme originally present is found from a calibration graph. The reproducibility of these graphs depends on the sample history, electrode history, reaction temperature, reaction time, pH and the presence of extraneous cations.

The potential developed (E) at the electrode surface is proportional to the activity of the ion $[A^{\pm}]$ to be measured as a result of the enzymatic reaction [21].

$$E = \text{Constant} + RT/nF \ln [A^{\pm}] \tag{4}$$

Differentiation of this equation with time gives equation 5 which forms the basis for the analysis of enzymes.

$$dE/dt = RT/2 \cdot 303 \, nF \times 1/[A^{\pm}] \times d[A^{\pm}]/dt \tag{5}$$

The concentration of the substrate is an important factor affecting the rate of a given enzymatic reaction. The correlation between enzyme activity and substrate concentration is given by the Michaelis–Menten equation [22]. This effect is shown in Fig. 3.1 for the response of urease electrode

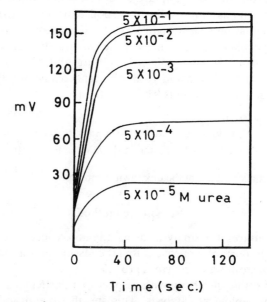

FIG. 3.1. Response curves for a urease electrode as a function of urea substrate concentration: 175 mg of urease/cm^3 of gel with a 350 μm netting is used for the preparation of the electrode.

as a function of urea (substrate) concentration. As the concentration of urea increases by a factor of 10, the steady state response increases until at high substrate concentration the steady state response is independent of the substrate concentration [23].

The standard calibration graphs for the enzymes are constructed by following one of the following methods.

1. Initial Slope Method

Using this method a tangent is drawn to the kinetic curves at the initial stage of the reaction (Fig. 3.2) and a calibration graph is constructed by plotting the slope $\Delta E/\Delta T$ as a function of enzyme activity [21, 24] (Fig. 3.3). These graphs show a linear relation for both enzymes and substrates with a slope not necessarily amounting to the Nernstian value.

From equation 5, when the reciprocal of the substrate is changed relatively more slowly with time, the time differential of the substrate in the initial stage of the reaction dE/dt is considered to be directly proportional to the rate of change in the potential $d[A^{\pm}]/dt$. With this method equation 5 holds for this early part of the reaction.

2. Lapsed Time Slope Method

This method eliminates the initial part of the curve in determining the rate of potential change. With this method, tangents are drawn to the enzymatic progress curves at a constant time after mixing the substrate and enzyme [21].

Construction of the calibration curves by this procedure (Figs 3.4, 3.5) resulted in a considerable diminishing of the sensitivity and applicable range compared to the initial slope method. However, this procedure offers the advantage of being applicable to almost any sample, regardless of its history, and can be used for purity determination as in preparative studies.

3. Time Potential Method

Using this method, the potential is measured after a constant time, and plotted as a function of enzyme activity (Figs 3.4, 3.5). The suitable time is chosen at which the reaction rate is linear. This method has the same disadvantages as the initial slope method and, therefore, is not recommended as a general procedure for preparing calibration graphs [21], but can be used, for example, to follow changes in enzyme activity due to denaturation when working with samples having the same history.

This method is also used for the determination of other species in non-enzymatic reactions. Proteins and protein mixtures in serum are determined [25] by reaction with silver, and measuring the potential of the free silver as a function of time. The potential readings are taken at a fixed time after the immersion of the electrode in the reaction solution and the concentration–potential graph is constructed.

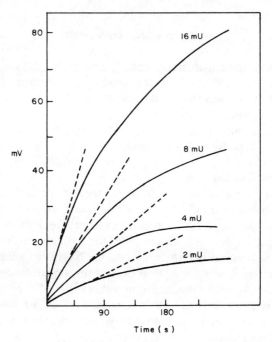

FIG. 3.2. Response curves of potential *vs* time for urease catalyzed hydrolysis of urea. Urea concentration 0·01 M, pH 7. (Courtesy of The American Chemical Society.)

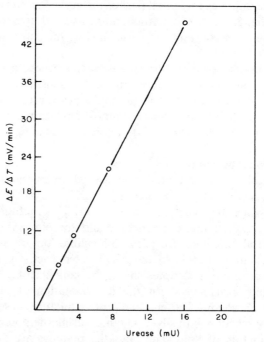

FIG. 3.3. Urease calibration graph using the initial slope method and the data given in Fig. 3.2.

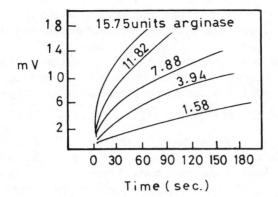

FIG. 3.4. Response curves for arginase by the urease electrode. (Courtesy of The American Chemical Society.)

4. *Automated Method*

Automatic analysis based on fixed kinetic potentiometric measurements using the ion-selective electrodes has been described for glucose [26] and enzymes [27]. This is performed by continuous recording of the potential output of the electrode after a fixed time. The reading for each sample is displayed as a peak on the recorder chart, so that a series of well defined peaks is obtained, one for each sample in sequence. The height of the peak is related, through the Nernst equation, to the concentration of the measured ion, and *via* a calibration graph to the activity of the species to be determined. The shapes of the peaks depend upon the response time

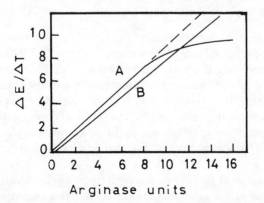

FIG. 3.5. Arginase calibration graphs by lapsed time slope and timed-potential methods using the data given in Fig. 3.4. (A) data points were obtained by constructing tangents to the curve after the initial 30 s; (B) data points were obtained by taking the potential after a constant time of 30 s (according to Booker and Haslam, *Anal. Chem.* **46**, 1054 (1974)). (Courtesy of The American Chemical Society.)

of the electrode, the fraction of sample sensed, the reaction time interval, the temperature, the nature of the reaction, and the efficiency of the washing cycle. These variables must be adjusted in the analysis.

On this basis, many inorganic and organic substances can be determined through specific catalytic reactions. For example, molybdenum and tungsten are known to catalyze the reaction of iodide with hydrogen peroxide:

$$H_2O_2 + 3I^- + 2H^+ \rightarrow 2H_2O + I_3^- \tag{6}$$

The decrease of iodide ion activity with time can be followed using the iodide ion-selective electrode and iodide activity vs time curves are constructed. The slopes of the linear parts of these curves are proportional to the concentration of molybdenum and tungsten [28]. Amounts as low as 0.004 $\mu g/ml$ of these catalysts can be determined. This level is lower than that obtained by other analytical procedures such as the colorimetric or chronometric methods.

III. STANDARD ADDITION AND SUBTRACTION METHODS

A. Known addition "spiking" technique

This technique involves the measurement of the potential of the unknown solution before and after the addition of one aliquot containing a known concentration of the test ion [29, 30]. The change in the e.m.f. can be expressed by the Nernst equation:

$$E_1 = \text{Constant} + S \log (\gamma_1 f_1 . C_0) \tag{7}$$

$$E_2 = \text{Constant} + S \log [\gamma_2 f_2 . (C_0 + C_\Delta)] \tag{8}$$

where E_1 and E_2 are the potentials of the initial test solution and the spiked test solution, respectively, S is the electrode slope $(2.3RT/nF)$ in mV per decade change in activity, which is not necessarily the theoretical Nernst value, and must be confirmed experimentally by a series of known standard solutions covering the whole concentration range of measurements. γ is the activity coefficient of the ion, f is the mole fraction of the free ion, C_0 is the total concentration of the ion in the unknown sample, and C_Δ is the increase in the concentration of the test ion produced by the standard addition.

Since the change in total ionic strength upon addition can be neglected, and the fraction of the ion being measured remains essentially unchanged (i.e. $f_1 \sim f_2$ and $\gamma_1 \sim \gamma_2$):

$$\Delta E = S \log \frac{(C_0 + C_\Delta)}{C_0} \tag{9}$$

$$\frac{C_\Delta}{C_0} = 10^{\Delta E/S} - 1 \tag{10}$$

$$C_0 = C_\Delta (10^{\Delta E/S} - 1)^{-1} \tag{11}$$

$$C_0 = C_\Delta \frac{1}{(\text{antilog } \Delta E/S) - 1} \tag{12}$$

where

$$C_\Delta = \frac{V_s C_s}{V_0} \tag{13}$$

and C_s is the molar concentration of the standard solution whose volume is V_s added to the unknown sample whose volume is V_0.

If the volume of the standard V_s is very small in comparison with that of the original test solution, it can be neglected and hence equation 12 becomes:

$$C_0 = C_s \left(\frac{V_s}{V_0}\right)\left(\frac{1}{\text{antilog } \Delta E/S - 1}\right) \tag{14}$$

However, if the volume of the standard solution V_s is added in large amounts that cannot be neglected with respect to the unknown solution, the change in volume must be considered, and equation 14 becomes:

$$C_0 = C_s \left(\frac{V_s}{V_s + V_0}\right)\left[10^{\Delta E/S} - \left(\frac{V_0 + V_s}{V_0}\right)\right]^{-1} \tag{15}$$

Practically, however, a small volume of the standard solution (about one one-hundredth of the unknown sample) equivalent to about 100 times the expected concentration of the unknown is used; therefore equation 14 is to be applied. The calculation of C_0 in this equation is simplified by tabulating the change in potential ΔE versus $1/(\text{antilog } \Delta E/S - 1)$ or C_0/C_Δ. Computer print-out tables of appropriate C_Δ and ΔE values for monovalent and divalent ions may be used (see Table 3.1). Thus, the calculation of the concentration of the unknown solution simply requires measurement of the potential difference ΔE and using Table 3.1 for the corresponding C_0/C_Δ value. For example, if a lead solution (100 ml) shows a potential change ΔE of 15 mV after addition of 1 ml of 0·1 M standard lead solution, then from Table 3.1 the change in potential of 15 mV (C_0/C_Δ) is equal to 0·452 and consequently, according to equation 14 the concentration of the unknown lead solution is

$$C_0 = \frac{0·1 \times 1}{100} \times 0·452 = 0·452 \times 10^{-3} \text{ M}$$

Table 3.1. C_0/C_Δ values (i.e. $1/(antilog \Delta E/S - 1)$ as a function of the change in potential ΔE for mono- and divalent electrodes

$\Delta E_{mono}(mV)$	$\Delta E_{di}(mV)$	C_0/C_Δ	$\Delta E_{mono}(mV)$	$\Delta E_{di}(mV)$	C_0/C_Δ
1	0·5	25·20	22	11	0·74
2	1·0	12·35	24	12	0·65
3	1·5	8·07	26	13	0·57
4	2·0	5·95	28	14	0·51
5	2·5	4·65	30	15	0·45
6	3·0	3·80	32	16	0·40
7	3·5	3·19	34	17	0·36
8	4·0	2·74	36	18	0·33
9	4·5	2·38	38	19	0·30
10	5·0	2·10	40	20	0·27
11	5·5	1·87	42	21	0·24
12	6·0	1·68	44	22	0·22
13	6·5	1·51	46	23	0·20
14	7·0	1·38	48	24	0·18
15	7·5	1·24	50	25	0·17
16	8·0	1·16	52	26	0·15
17	8·5	1·07	54	27	0·14
18	9·0	0·99	56	28	0·13
19	9·5	0·92	58	29	0·12
20	10·0	0·85	60	30	0·11

The calculation may also be simplified by the use of the Orion Specific Ion Meter (Model 407) which contains a scale indicating the ratio of the amount of ion C_0 in the original solution to the amount added in the known increment C_s. Simply, the test solution is completed to 100 ml and the function switch of the ion meter is set at the ion species position (monovalent or divalent), the slope indicator is set at 100% and the temperature compensator at the temperature of the solution. The calibration knob is turned to the centre of the scale, and 1 ml of the known standard solution is added. The reading is recorded on the green known addition scale. The original concentration C_0 is obtained by multiplying the meter reading by the change in concentration C_Δ (equation 13).

The growing interest on the application of this technique [31–45] is due to its advantages which can be seen in (i) the measurement of the total concentration of the complexed or uncomplexed ions, even in very complicated systems, and (ii) the measurement of ions in solutions having widely varying ionic strength. This method is the only one available for the determination of the total concentration of the complexed species using the ion-selective electrodes.

B. Known addition–dilution technique

This technique involves the addition of an aliquot of known concentration (to make the unknown concentration approximately double) and measuring the potential change ΔE_a. The sample is then diluted from the initial volume V_i to a final volume V_f, and the change in potential caused by dilution (E_{dil}) is measured [46]. The electrode slope is:

$$S = \frac{\Delta E_{dil}}{\log (V_f/V_i)} \tag{16}$$

For most samples, it is convenient to make $V_f = 2V_i$, thus equation 16 takes the form:

$$S = \frac{\Delta E_{dil}}{0\cdot301} \tag{17}$$

Practically, the potential of a 100 ml sample solution (ΔE_0) is measured before and after the addition of 1 ml of the known standard solution (about 100 times more concentrated). Then 50 ml of the sample is withdrawn by a pipette and replaced with 50 ml of distilled water. The potential is measured again and ΔE_{dil} is estimated. Suppose that ΔE_a is 20 mV using 10^{-5} M of the standard solution, and ΔE_{dil} is 10 mV at 1:1 dilution. Then the value C_0/C_Δ is obtained by calculating the ratio of $\Delta E_a/\Delta E_{dil}(R)$ and using Table 3.2. In the above-mentioned example, $R = 2$ which corresponds to the value $C_0/C_\Delta = 0\cdot333$. Thus, the concentration of the unknown amounts to:

$$C_0 = \frac{1 \times 10^{-5}}{100} \times 0\cdot333 = 3\cdot33 \times 10^{-6} \text{ M}$$

Dilution with water is usually employed for pure solutions, but in the presence of a constant level of an interfering ion or complexing agent, the levels of these substances must remain constant, not only when the addition is made but also when the sample is diluted. In the presence of an interfering ion, a suitable background solution can be used, whereas in the presence of complexing species, a diluent of a decomplexing agent with water in the same ratio is to be used as is the case with the sample.

This method is useful because it is applicable to samples of very low concentrations, or in the range beyond the specified operating range of the electrode, or when the electrode response is not Nernstian. By means of this technique fluoride concentrations within 10^{-8} M or p.p.b. levels (i.e. two orders of magnitude below the specified lower limit of detection) can be measured [46].

Table 3.2. C_0/C_Δ values as a function of $\Delta E_a/\Delta E_{dil}(R)$ using 1 : 1 dilution

$\Delta E_a/\Delta E_{dil}$	C_0/C_Δ	$\Delta E_a/\Delta E_{dil}$	C_0/C_Δ	$\Delta E_a/\Delta E_{dil}$	C_0/C_Δ
0·20	6·727	0·90	1·155	1·60	0·492
0·22	6·072	0·92	1·121	1·62	0·482
0·24	5·526	0·94	1·089	1·64	0·473
0·26	5·066	0·96	1·058	1·66	0·463
0·28	4·670	0·98	1·028	1·68	0·454
0·30	4·327	1·00	1·000	1·70	0·445
0·32	4·026	1·02	0·973	1·72	0·436
0·34	3·763	1·04	0·947	1·74	0·427
0·36	3·528	1·06	0·922	1·76	0·419
0·38	3·319	1·08	0·898	1·78	0·411
0·40	3·130	1·10	0·875	1·80	0·403
0·42	2·959	1·12	0·852	1·82	0·395
0·44	2·805	1·14	0·831	1·84	0·388
0·46	2·663	1·16	0·810	1·86	0·380
0·48	2·533	1·18	0·790	1·88	0·373
0·50	2·415	1·20	0·771	1·90	0·366
0·52	2·305	1·22	0·752	1·92	0·359
0·54	2·203	1·24	0·734	1·94	0·353
0·56	2·109	1·26	0·717	1·96	0·346
0·58	2·021	1·28	0·700	1·98	0·340
0·60	1·939	1·30	0·684	2·00	0·333
0·62	1·863	1·32	0·668	2·02	0·327
0·64	1·791	1·34	0·653	2·04	0·321
0·66	1·724	1·36	0·638	2·06	0·316
0·68	1·661	1·38	0·624	2·08	0·310
0·70	1·602	1·40	0·610	2·10	0·304
0·72	1·542	1·42	0·597	2·12	0·299
0·74	1·492	1·44	0·584	2·14	0·294
0·76	1·442	1·46	0·571	2·16	0·288
0·78	1·394	1·48	0·559	2·18	0·283
0·80	1·349	1·50	0·547	2·20	0·278
0·82	1·307	1·52	0·535	2·22	0·273
0·84	1·266	1·54	0·524	2·24	0·269
0·86	1·227	1·56	0·513	2·26	0·264
0·88	1·190	1·58	0·503	2·28	0·259

C. Analate addition potentiometric technique

Using this technique, an aliquot of the unknown (analate) solution is added
to a known volume of a standard solution containing the sample species,
and the change in e.m.f. is then related to the unknown concentration [47,
48].

$$C_0 = C_s\left[\left(\frac{V_s + V_0}{V_0}\right)10^{\Delta E/S} - \left(\frac{V_s}{V_0}\right)\right] \tag{18}$$

where C_0 and V_0 and C_s and V_s are the concentrations and volumes of the unknown and standard samples, respectively.

If $K = (V_s + V_0)/V_0$, then:

$$C_0 = C_s[K \cdot 10^{\Delta E/S} - (K - 1)] \tag{19}$$

This technique is used when the volume of the unknown sample is small. Thus the volume suitable for electrode immersion can be satisfied by the standard solution, and the unknown (analate) aliquot is added to develop a measurable potential change. As the concentration and volume of the standard sample decreases, the required volume of the unknown sample is reduced. Since the background of the unknown sample solution differs from the standard, an error may be introduced when the two solutions are mixed. This is only true when the degree of complexation changes by mixing the solutions. The precision of this method is about 0·5% with a 95% confidence limit.

D. Multiple addition technique

With this technique, two or more successive aliquots of equal concentrations of the standard solution containing the ion to be measured are added to a known volume of the sample solution, and the electrode potentials (E_2 and E_3) are measured after each addition [46].

$$E_2 = \text{Constant} + S \log f_2\gamma_2(C_0 + C_a) \tag{20}$$

$$E_3 = \text{Constant} + S \log f_3\gamma_3(C_0 + C_a + C_a) \tag{21}$$

where C_a is the change in concentration of the sample solution on addition of the standard; it can be made equal to one (i.e. 1 p.p.m., 1 mEq/L or 1×10^x M) in order to simplify the calculation. At constant f and γ:

$$\Delta E_2 = E_2 - E_1 = S \log \frac{C_0 + C_a}{C_0} \tag{22}$$

$$\Delta E_3 = E_3 - E_1 = S \log \frac{C_0 + 2C_a}{C_0} \tag{23}$$

$$R = \frac{\Delta E_3}{\Delta E_2} = \frac{\log \dfrac{C_0 + 2C_a}{C_0}}{\log \dfrac{C_0 + C_a}{C_0}} \tag{24}$$

Table 3.3 is constructed for R versus C_0 by successive approximations on a computer, and is used for the calculation of the sample concentrations.

Table 3.3. C_0/C_Δ values as a function of $\Delta E_3/\Delta E_2(R)$

R	C_0/C_Δ	R	C_0/C_Δ	R	C_0/C_Δ
1·270	0·100	1·520	0·694	1·670	1·598
1·280	0·113	1·525	0·714	1·675	1·643
1·290	0·126	1·530	0·735	1·680	1·691
1·300	0·140	1·535	0·756	1·685	1·738
1·310	0·154	1·540	0·778	1·690	1·787
1·320	0·170	1·545	0·801	1·695	1·840
1·330	0·186	1·500	0·823	1·700	1·894
1·340	0·203	1·555	0·847	1·705	1·948
1·350	0·221	1·560	0·870	1·710	2·006
1·360	0·240	1·565	0·896	1·715	2·066
1·370	0·260	1·570	0·920	1·720	2·126
1·380	0·280	1·575	0·946	1·725	2·190
1·390	0·302	1·580	0·973	1·730	2·256
1·400	0·325	1·585	1·000	1·735	2·326
1·410	0·349	1·590	1·029	1·740	2·397
1·420	0·373	1·595	1·056	1·745	2·470
1·430	0·399	1·600	1·086	1·750	2·549
1·400	0·427	1·605	1·116	1·755	2·629
1·450	0·455	1·610	1·147	1·760	2·711
1·460	0·485	1·615	1·179	1·765	2·801
1·470	0·516	1·620	1·213	1·770	2·892
1·475	0·532	1·625	1·245	1·775	2·985
1·480	0·548	1·630	1·280	1·780	3·088
1·485	0·565	1·635	1·315	1·785	3·193
1·490	0·582	1·640	1·353	1·790	3·301
1·495	0·600	1·645	1·391	1·795	3·416
1·500	0·618	1·650	1·430	1·800	3·536
1·505	0·637	1·655	1·469	1·805	3·664
1·510	0·655	1·660	1·510	1·810	3·797
1·515	0·675	1·665	1·554	1·815	3·939

For example, if 100 ml of the unknown sample solution shows ΔE_2 of 20 mV and ΔE_3 of 30 mV by addition of 1 ml containing 1 p.p.m. of the ion to be measured, then R is equal to 1·50. By using Table 3.3, the C_0 value is found to be equal to 0·618 and the concentration C_0 of the original solution is then 0·618 p.p.m.

This method offers the advantage that the electrode slope, which may be different from the theoretical, especially near the lower limit of detection or in regions in which the electrode interferences are appreciable, need not be known. Under ideal conditions, the accuracy of the double addition technique is between five and ten times worse than other electrode methods. An error within $0\cdot2$ mV for ΔE gives rise to an error of 8% in concentration.

A computer program (ADDFIT in Fortran IV) has been described [49] for use with double and multiple standard addition techniques based on equation 25.

$$E_i = E_0 + S \log \frac{C_0 V_0 + \sum_{i=1}^{j} C_s V_s}{V_0 + \sum_{i=1}^{j} V_s} \tag{25}$$

This is applicable for any number $(j-1)$ of additions of V_s ml of a solution of known concentration C_s to V_0 ml of solution of unknown concentration C_0. In case of multiple additions, the system of equations is over-defined, and it is necessary to select the best values of E_0, S and C to fit the data to the equation. A curve-fitting technique involving three parameters has been proposed [49a]; results can be processed on a programmable desk calculator.

E. Known addition, multiple addition and analate addition techniques using a standard nomograph

By using the nomograph constructed by Karlberg [50] (Fig. 3.6), the concentration of the mono- and divalent ions in the sample solution can be read off after applying one of the techniques described above.

1. Known Addition Technique

Suppose that the initial volume of the unknown sample (V_0) is 40 ml, the volume of the standard solution (V_s), whose concentration is 10^{-2} M, is 5 ml, the slope of the electrode (S) is $58\cdot5$ mV, and the potential difference caused by the addition (ΔE) is $34\cdot3$ mV. Then (i) the ratio of $V_0/V_s(40/5 = 8)$ together with $\Delta E(34\cdot3$ mV) gives point (a) in Fig. 3.6; (ii) by following the guideline from point (a) to point (b), the horizontal line at the value of the slope electrode ($58\cdot5$ mV) intersects with a vertical line from point (b) to give point (c); (iii) following the guideline from point (c) to point (d) and the value of the log scale (F) is read, and C_s is divided by this value to give the concentration of the sample: $C_0 = 10^{-2}/27 = 3\cdot7 \times 10^{-4}$ M. Calculation of the sample concentration (C_0) using equation 26 gives a value of $3\cdot74 \times 10^{-4}$ M.

$$C_0 = \frac{C_s}{10^{E/S}(1 + V_0/V_s) - V_0/V_s} \tag{26}$$

2. Multiple Addition Technique

In the above-mentioned example, if another 5 ml of the standard solution are added to the unknown sample $(V_s = 5 + 5 = 10$ ml), and the change in potential caused by the second addition is $45\cdot0$ mV, then the V_0/V_s ratio

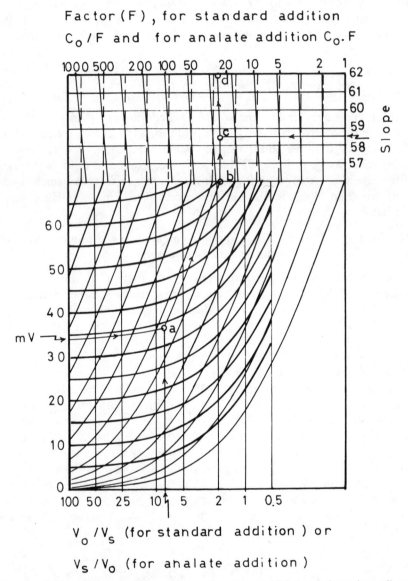

FIG. 3.6. Nomograph for standard addition and analate addition techniques (according to Karlberg, *Anal. Chem.* **43**, 1911 (1971)). (Courtesy of The American Chemical Society.)

amounts to $40/10 = 4$. Following the above steps (i–iii) the value obtained for C_0 is the same as that obtained by the single known addition technique.

3. Analate Addition Technique

As an example, suppose that the volume of the standard solution (V_s) is 20 ml, its concentration is 10^{-4} M, the volume of the unknown sample added (V_0) is 2·5 ml, the slope (S) is $-58·5$ mV, and the change in potential is 34·3 mV. To read the concentration of the unknown (C_0): (i) follow the guideline from point (a) up to point (b); (ii) the values of (S) and a vertical line from point (b) intersect to give point (c); (iii) follow the guideline from point (c) to point (d), read the value of the log scale (F) and multiply by C_s. Therefore C_0 is $10^{-4} \times 27 = 2·7 \times 10^{-3}$ M. Calculation of this value using equation 27 gives a value of $2·67 \times 10^{-3}$ M. Consequently, the difference between the results obtained using the nomograph, and those obtained by calculation is within $\pm 1\%$.

Application of the nomograph to measure the divalent ions is also possible by reducing the slope scale, and the millivolt scale in the nomograph to their half values, using all the other parameters without change:

$$C_0 = C_s[10^{E/S}(1 + V_s/V_0) - V_s/V_0] \tag{27}$$

F. Known subtraction technique

This technique is preferred to the known addition technique when unstable ions (e.g. sulfide ions) are to be determined. The method is based on the measurement of the potential developed in 100 ml of the sample solution before (E_1) and after (E_2) addition of 1 ml of a standard solution containing a suitable precipitating or complexing agent in sufficient quantity to precipitate or complex about half or all the amount of the ion to be measured [30, 31]. The $E_2 - E_1$ value will be negative since $E_2 < E_1$. Then:

$$-\Delta E = S \log \frac{C_0 - C_\Delta}{C_0} \tag{28}$$

Consequently equation 12 becomes:

$$C_0 = C_\Delta \left[\frac{1}{(1/\text{antilog } \Delta E/S) - 1} \right] \tag{29}$$

$$C_0 = C_s \left(\frac{V_s}{V_0} \right) \left[\frac{1}{(1/\text{antilog } \Delta E/S) - 1} \right] \tag{30}$$

Equation 30 is valid only when the valency of the unknown ion and that of the precipitating or complexing agent added is the same, and their stoichiometric reaction ratio is $1:1$. When the valency differs, the stoichiometry differs and the term C_Δ must be corrected. The term $C_s V_s/V_0$

is multiplied by a factor of 2 when the standard reagent is divalent and the unknown ion is monovalent, and divided by a factor of 2 when the standard reagent is monovalent and the unknown ion is divalent.

For example, if the decrease in potential reading caused by addition of 1 ml of 10^{-1} M silver nitrate (as a precipitating agent) to 100 ml of unknown sulfide solution is 15 mV, the concentration of sulfide can be calculated as follows:

$$C_\Delta = 1/2(C_s V_s / V_0) = 1/2 \cdot 10^{-1}/100 = 5 \times 10^{-4} \text{ M}$$

The factor $1/2$ is used since each sulfide ion reacts with 2 ions of silver. The C_0/C_Δ value (equation 30) can be read off directly from the print-out computer (Table 3.4). The value equivalent to ΔE of 15 mV in the above mentioned example is 1·45, therefore $C_0 = 5 \times 10^{-4} \times 1 \cdot 45 = 7 \cdot 25 \times 10^{-4}$ M.

The sulfide ion is also determined by adding an excess of cadmium ion and measurement of the loss of cadmium, using the cadmium ion-selective electrode [51]. Unstable oxidizing agents such as active chlorine, bromine, hypochlorite, hypobromite and chromium (VI) are determined by the addition of excess iodide ion, and measuring the loss in iodide using the iodide ion-selective electrode [52].

Table 3.4. C_0/C_Δ values (i.e. $1/(1/\text{antilog } \Delta E/S - 1)$) as a function of the change in potential ΔE for mono- and divalent electrodes

ΔE_{mono}(mV)	ΔE_{di}(mV)	C_0/C_Δ	ΔE_{mono}(mV)	ΔE_{di}(mV)	C_0/C_Δ
1	0·5	25·00	22	11	1·74
2	1·0	13·35	24	12	1·65
3	1·5	9·07	26	13	1·57
4	2·0	6·95	28	14	1·51
5	2·5	5·65	30	15	1·45
6	3·0	4·80	32	16	1·40
7	3·5	4·19	34	17	1·36
8	4·0	3·74	36	18	1·33
9	4·5	3·38	38	19	1·30
10	5·0	3·10	40	20	1·27
11	5·5	2·87	42	21	1·24
12	6·0	2·68	44	22	1·22
13	6·5	2·51	46	23	1·20
14	7·0	2·38	48	24	1·18
15	7·5	2·24	50	25	1·17
16	8·0	2·16	52	26	1·15
17	8·5	2·07	54	27	1·14
18	9·0	1·99	56	28	1·13
19	9·5	1·92	58	29	1·12
20	10·0	1·85	60	30	1·11

IV. POTENTIOMETRIC TITRATION METHODS

The potentiometric titration technique offers the advantages of high accuracy and precision, applicability to aqueous and non-aqueous systems, and the possibility of using a selective ion as a titrant in conjunction with its electrode for the determination of other species, for which ion-selective electrodes are unsuitable or not available. Tables 3.5 and 3.6 show the usefulness of the ion-selective electrodes in many complexometric and precipitation titrations of various cations and anions.

A. Titration curves

All titration procedures involve measurement and recording of the cell potential after each addition of the titrant. At the beginning of the titration, the titrant is gradually added in large increments and thus the concentration of the ion to be measured decreases with a slow change in the electrode potential. At the end point, all the free ion has reacted. Further addition of a few drops of the titrant after the end point causes an increase in the concentration of the titrant with a corresponding decrease in the free ion, since the concentration of these ions are inversely proportional to the concentration of the titrant. This sudden decrease in ion concentration at the end point is reflected in the abrupt change in electrode potential.

The equivalence point is calculated from the mid-point in the steeply rising portion of the sigmoid titration curves. Symmetrical titration curves are obtained when the reactants undergo a reaction in an equimolar ratio, provided that the electrode process is perfectly reversible. If these conditions are not met, asymmetric titration curves are obtained.

A plot of the change in potential per unit change in volume of reagent ($\Delta E/\Delta V$) as a function of the average volume of reagent added may also be used. The end point is taken as a maximum in the curve, and is obtained by extrapolation of the experimental points. Using second derivative data, a less time consuming method is applied for calculation of the equivalence point from the point where $\Delta^2 E/\Delta V^2$ becomes zero (i.e. corresponding to the maximum of the first derivative). This method does not require graphical evaluation and is less prone to personal error.

During the titration, sufficient time must be allowed for the attainment of equilibrium after each addition of the reagent. Precipitation titrations may require several minutes for equilibrium, particularly in the vicinity of the equivalence point as indicated by potential jumps. Efficient stirring is frequently effective in hastening the achievement of these conditions.

1. Complexometric and Precipitation Titrations

The ion-selective electrodes are usually used as detectors for potentiometric titrations involving precipitation and complexometric reactions. The

Table 3.5. Complexometric titration of some ions using ion-selective electrodes

Ion determined	Titrant	Ion-selective electrode	References
Aluminum (+EDTA)	Cu^{2+}	Copper	97
(+CuEDTA)	CDTA	Copper	97
	F^-	Fluoride	98–101
Barium	EDTA	Calcium	102
Bismuth (III) (+HgEDTA)	EDTA	Mercury	103
(+MgEDTA)	EDTA	Divalent	104
Cadmium	EDTA, CDTA	Cadmium	82, 105–107
Calcium	EDTA	Calcium	102, 108–112
		Copper	113
Cerium (+MgEDTA)	EDTA	Divalent	104
Chromium (+MgEDTA)	EDTA	Divalent	104
(+HgEDTA)	EDTA	Mercury	103
Citrate	Cu^{2+}	Copper	114
Cobalt (CuEDTA)	EDTA	Copper	114
Copper	EDTA	Copper	105, 113, 115
Cyanide	Ni^{2+}	Silver/sulfide	116
Divalent Cations	TTHA	Calcium	117
Gallium (+MgEDTA)	EDTA	Divalent	104
Indium (+MgEDTA)	EDTA	Divalent	104
Iron (III) (+CuEDTA)	EDTA	Copper	113
Lanthanum (III) (+CuEDTA)	EDTA	Copper	113
Magnesium	EDTA	Calcium	108, 112
Manganese (II) (+CuEDTA)	EDTA	Copper	114
Mercury (II) (+CuEDTA)	EDTA	Copper	113
Nickel (+CuEDTA)	TEPA, EDTA	Copper	57
(+CdEDTA)	EDTA	Cadmium	118
Nitrilotriacetic acid	Cu^{2+}	Copper	119
Samarium (III) (+CuEDTA)	EDTA	Copper	113
Thallium (+MgEDTA)	EDTA	Divalent	104
Thorium (IV) (+CuEDTA)	EDTA	Copper	113
(+MgEDTA)	EDTA	Divalent	104
Trivalent Cations	TTHA	Calcium	117
Uranium (+MgEDTA)	EDTA	Divalent	104
Vanadium (+MgEDTA)	EDTA	Divalent	104
Zinc (+CuEDTA)	TEPA, EDTA	Copper	57
Zirconium (+CuEDTA)	EDTA	Copper	113
(+MgEDTA)	EDTA	Divalent	104

EDTA = Ethylene Diamine Tetra Acetic acid disodium salt.
TEPA = Tetra Ethylene Penta Amine.
CDTA = 1,2-Cyclohexylene Diamine Tetra Acetic acid.
TTHA = Triethylene Tetraamine Hexa Acetic Acid.

Table 3.6. Precipitation titration of some ions using ion-selective electrodes

Ion determined	Titrant	Ion-selective electrode	References
Ammonium	Calcium tetraphenyl boron	Cation	120
Arsenate ($+La^{3+}$)	F^-	Fluoride	121
Azide	Ag^+	Chloride	64
		Silver	122
Barium	SO_4^{2-}	Barium	123
Cadmium	S^{2-}	Cadmium	124
Caesium	Calcium tetraphenyl boron	Cation	120
Cyanide	Ag^+	Silver/sulfide	125, 126
		Cyanide	126
		Silver	127
Fluoride	La^{3+}	Fluoride	102, 128–133
	Th^{4+}	Fluoride	130, 134, 135
	Ca^{2+}	Fluoride	130
	Tetraphenyl antimony sulfate	Fluoride	136
Halides (Cl, Br, I)	Ag^+	Chloride	137–142
	Ag^+	Iodide	58, 143, 144
	Ag^+	Silver/sulfide	102, 145, 146
	Pb^{2+}	Silver/sulfide	147
	Hg^{2+}	Silver/sulfide	148, 149
Lead	Sodium molybdate or tungstate	Lead	125
Lithium	Ammonium fluoride	Fluoride	150
Mercuric	I^-	Iodide	151
Mercurous	Br^- or I^-	Iodide	63, 152
Molybdate	Pb^{2+}	Lead	125
Nitrate	Diphenyl thallium (III) sulfate	Nitrate	153
Oxalate	Pb^{2+}	Lead	154
Perbromate	Tetraphenyl arsonium chloride	Perchlorate	155
Perchlorate	Tetraphenyl arsonium chloride	Perchlorate	156
	Hg^{2+}	Chloride	157
Periodate	Tetraphenyl arsonium chloride	Perchlorate	155
Phosphate	Pb^{2+}	Lead	158
	Ag^+	Silver/sulfide	159
($+La^{3+}$)	F^-	Fluoride	113
Phosphonic	Ca^{2+}	Divalent	160
Potassium	Calcium tetraphenyl boron	Cation	120, 161
	Zn (SiF_6)	Glass	162

Table 3.6. (cont.)

Ion determined	Titrant	Ion-selective electrode	References
Rubidium	Calcium tetraphenyl boron	Cation	120
Silver	MgCl$_2$	Glass	163
	CN$^-$	Silver	164
	Calcium tetraphenyl boron	Cation	120
Strontium	SO$_4^{2-}$	Lead	165
Sulfate	Ba^{2+}	Iron	166
	Pb^{2+}	Lead	167–174
Sulfide	Cd^{2+}	Silver/sulfide	175
	Sodium plumbite	Silver/sulfide	176, 177
	Ag$^+$	Silver/sulfide	178
Tetrafluoroborate	Tetraphenyl arsonium chloride	Fluoroborate	179
Tetraphenyl boron	Ag$^+$	Halide	180
Thiocyanate	Ag$^+$	Silver/sulfide	102
	Pb^{2+}	Lead	147
Tungstate	Pb^{2+}	Lead	125

magnitude of the break at the end point affects the accuracy and precision of the results. This depends on both the concentration of the ion to be determined and the stability constant of the complex (β) in a complexometric titration or the solubility product (K_{sp}) of the precipitate in a precipitation titration [53, 54].

In complexometric titrations, as the stability of the complex and concentration of the ion increase, the magnitude of the break increases. The presence of organic radicals such as citrate, tartrate or strong acids limits the break, since these species compete with the metal. In precipitation titrations, co-precipitation and formation of solid solutions in the precipitate phase affect the inflection points. However, great improvements can be obtained if water-miscible organic solvents (methanol, acetone, dioxane) are added to the solution prior to titration in amounts ranging from 20 to 80% [55, 56]. These solvents decrease the dielectric constant of the solution and reduce the solubility product of the precipitate.

The potential difference between 50% and 150% of reaction is taken as a measure of the magnitude of the break, and can be calculated according to equations 32 and 34. For example, in complexometric titrations [53]:

$$M + L \rightleftharpoons ML; \beta = ML/[M][L] \tag{31}$$

$$E = RT/n\text{F} \log \bar{M}\beta/4 \tag{32}$$

For precipitation titrations [54]:

$$nX + yP \rightleftharpoons X_nP_y(S); \quad K_{sp} = (X)^n(P)^y \tag{33}$$

$$E = RT/nF \log \frac{y^{y/n}/n \cdot \bar{X}^{1+y/n}/2}{K^{1/n}} \tag{34}$$

where $\bar{M}$ and $\bar{X}$ are the initial concentrations of the ion to be titrated, β is the stability constant, K is the solubility product, and y and n are the numbers of ions involved in the reaction.

The end point break can be calculated by knowing the strength of the solutions and the solubility product or stability constant. These two constants can also be calculated for solutions of known strength. For example, an end point break of 60 mV is obtained for the titration of a 10^{-3} M divalent metal ion with a titrant capable of forming a complex having a β value of at least 4×10^5, whereas the same break is obtained for the precipitation titration of a monovalent ion $(n = y = 1)$ if the solution to be titrated is 10^{-2} M and the K_{sp} is $2 \cdot 5 \times 10^{-5}$, or with a solution of 10^{-4} M and K_{sp} of $2 \cdot 5 \times 10^{-9}$, or 10^{-6} M solution and K_{sp} $2 \cdot 5 \times 10^{-13}$.

2. Sequential Complexometric Titrations

Simultaneous titration of two ions or more with the same titrant is possible if an electrode sensing the ion which forms the more stable complex or the less soluble precipitate is available [181–185].

Complexometric titrations of two metal ions (M_1 and M_2) using the ion-selective electrodes show that the ion which forms the more stable complex (M_1) is titrated first; after all the M_1 has been complexed by the ligand (L), the concentration of the latter increases up to a point where

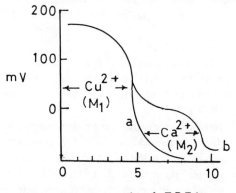

FIG. 3.7. Potentiometric titration of copper (II) with the copper electrode. (a) Titration of pure copper solution with EDTA; (b) titration of copper–calcium mixture with EDTA. (Courtesy of Orion Research Inc.)

the second complex M_2L begins to form. Addition of the ligand does not cause an appreciable decrease in the concentration of M_1, and consequently the break is less than when M_1 is present alone (Fig. 3.7). The magnitude of the second end point break depends on the initial concentration of $\bar{M}_2$, and the formation constant β_2 ($\beta_2 < \beta_1$) for the complex M_2L, but is independent of the initial concentration of $\bar{M}_1$ and can be calculated from equation 35.

$$E = RT/n\mathrm{F} \log [\beta_2 \bar{M}_2/2] \tag{35}$$

The end point breaks and the optimum pH for a complexometric titration can be graphically estimated [57]. The potential of a fixed concentration of the metal M_1 in pure form, and in the presence of excess complexing titrant are measured as a function of pH (Fig. 3.8, curves i and ii). Then the potentials of a known concentration of the second metal ion (M_2) with exactly half the equivalent amount of the complexing titrant are measured as a function of the pH (Fig. 3.8, curve iii). The maximum end point break can be obtained at the pH for which the vertical distance between curves i, ii and iii is maximum. Figure 3.9 shows the possible simultaneous titration of copper and zinc ions with a maximum end point break at pH 10 using tetraethylene pentamine (TEPTA) as a titrant and the copper ion-selective electrode.

The actual potential-pH curves for copper-1,10 phenanthroline, tetraethylene pentamine-copper and copper-ethylene diamine tetraacetic acid (EDTA) indicate that zinc, nickel and cadmium ions can be titrated with 1,10-phenanthroline, lead, nickel and zinc with TEPTA; and nickel, cadmium and calcium with EDTA using the copper ion-selective electrode. Simultaneous titration of any of these metal ions with copper can also be performed by selecting the proper pH for the titration [57].

3. Sequential Precipitation Titrations

Potentiometric titration of halide mixtures using silver and mercurous nitrates, and the halide ion-selective electrodes has been described [58–63]. The solubility products and solubilities of silver and mercurous halides in water can be seen in Table 3.7.

By simple calculations using the above-mentioned solubility products, and the concentrations of the halides in water, one may expect that some of the bromide will be precipitated before the end point of iodide and some of the chloride will be precipitated before the end point of bromide. This is true for both silver and mercurous halides, but to different extents. Factors influencing the simultaneous determination of the halides by the ion-selective electrodes are adsorption, co-precipitation, and formation of a solid solution.

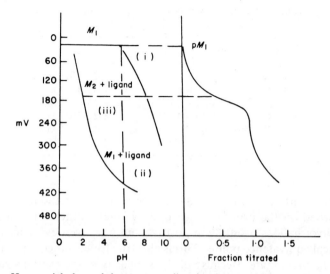

FIG. 3.8. pH-potential plot and the corresponding titration curve for a metal ion (according to Ross and Frant, *Anal. Chem.* **41**, 1900 (1969)). (Courtesy of The American Chemical Society.)

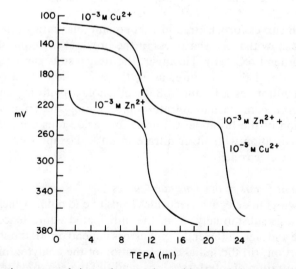

FIG. 3.9. Titration curves of zinc and copper ions with tetraethylene pentamine using the copper electrode (according to Ross and Frant, *Anal. Chem.* **41**, 1900 (1969)). (Courtesy of The American Chemical Society.)

Table 3.7. Solubility products and solubilities of silver
and mercurous halides in water

Halide	K_{sp}	S
AgCl	$1 \cdot 78 \times 10^{-10}$	$4 \cdot 2 \times 10^{-5}$ M
AgBr	$5 \cdot 25 \times 10^{-13}$	$7 \cdot 3 \times 10^{-7}$ M
AgI	$8 \cdot 31 \times 10^{-17}$	$9 \cdot 1 \times 10^{-9}$ M
Hg_2Cl_2	$1 \cdot 3 \times 10^{-18}$	$3 \cdot 2 \times 10^{-7}$ M
Hg_2Br_2	$5 \cdot 8 \times 10^{-23}$	$2 \cdot 4 \times 10^{-8}$ M
Hg_2I_2	$4 \cdot 5 \times 10^{-29}$	$2 \cdot 2 \times 10^{-10}$ M

Titration with silver nitrate indicates that both silver and halide ions can be adsorbed on the primary or secondary layer of the colloidal particles of the silver halides resulting in positive and negative deviations from the theoretical equivalence points. This is also true for co-precipitation of these ions in the precipitate. Because of the similarity of the crystal structure of AgCl and AgBr (cubic), they are expected to form solid solutions during the titration in the presence of chloride ions. Silver iodide has a different crystal structure and its formation of solid solutions with other halides is not likely to occur. Therefore, one may expect in solid solution formation only, positive deviation for bromide and negative deviation for chloride. This error can be eliminated by taking the Fajans–Paneth adsorption rule into account, and adding excess of potassium or barium nitrate to the sample.

Titration curves for Cl, Br, and I, in equal molar ratios using the halide electrodes show that the sharpness of the break at the equivalence points is in the order $I > Cl > Br$. However, the best results can be obtained by titration of the halide mixture, using the iodide ion-selective electrode, with silver nitrate as a titrant in 85% ethanol, or with mercurous nitrate in the presence of barium nitrate while heating the solution just after precipitation of the bromide. Chloride and azide ions are determined by sequential titration with silver nitrate in 80% dioxane using the chloride ion-selective electrode [64].

4. *End-point Errors in the Titration Curves*
Titration with ion-selective electrodes leads to end points which may differ, in some cases substantially, from the anticipated equivalence point. The magnitude and sign of the end-point errors depend on: (i) the stoichiometry of the reaction; (ii) the initial concentration of the analyte; (iii) the nature and concentration of interfering ions; and (iv) the electrode selectivity.

In isovalent precipitation reactions, the titration curves are symmetrical. By contrast, in complexometric titrations the curves are intrinsically asymmetric. This asymmetry causes inflection point errors to result regardless of the stability constant of the complex reaction, whereas in precipitation titration no error occurs in the isovalent case [65].

Dilution may also affect the accuracy of the titration. During the titration, the volume of the system changes perceptively because of the addition of the titrant. This is mandated by the fact that the dilution effect always acts to decrease the magnitude of the potential break [66]. The effect of dilution can be minimized by using a titrant concentration which is much larger than the analyte concentration.

The presence of interfering ions in either the titrant or with the analyte ultimately limits the magnitude of the end point break and concomitantly the precision of measurement [65]. This interference effect may distort the titration curves and causes the inflection point to occur at a point other than the equivalence point. High concentration of the interfering ions may severely flatten the curves and cause the inflection point to become indistinct.

In complexometric titration with ion-selective electrodes, large errors may be obtained whenever ionic interferences are present in the analyte solution. Consequently, high ionic strength buffers are used to maintain the pH or to stabilize the liquid junction potential. Fortunately, most complexometric titrations are performed with reagents whose formation constants are quite large [66, 67]. Consideration of the ion charge ratio becomes important only when the dilution strongly influences the titration error. When interfering ions are introduced with the titrant, the dilution factor and ion charge ratio become important variables. This error is larger when a divalent ion interference is added to a univalent ion sample than when a univalent interference is added to a divalent sample. The latter case is the normal circumstance in complexometric titrations [66, 67]. The error due to ion interferences may be circumvented by using titrants free from, or containing negligible amounts of, interfering ions and to standardize the titrant with solutions as similar in composition to the sample to be analyzed as possible.

The most important factor determining the extent to which the interfering ions affect the titration is the selectivity coefficient of the ion-selective electrodes. Single crystal membrane-type electrodes exhibit very high selectivity for the ion of interest. Homogeneous solid state membrane-type electrodes have selectivity coefficients which are sufficient to tolerate the effect of interferences to some extent. However, many commercially available liquid membrane and solid membrane ion-exchange electrodes display only moderate selectivity and hence the effect of the interference ions is significant [68].

B. Gran's plot

In 1952, Gran [69] worked out a method for locating the equivalence points in titrations based on drawing linear titration curves. Recently, this method has gained wide application with the ion-selective electrodes.

By plotting the volume of titrant consumed against the concentration of the ion to be determined and sensed by the electrode, a straight line is obtained. As the volume of titrant added increases, the concentration of the ion decreases linearly to approach zero at the equivalence point, assuming that the sample and titrant have a high and constant ionic strength and that the activity coefficient of the ion in question and the liquid junction potential will not change appreciably. Since at the end point the concentration of the ion being sensed is very small, and the equilibrium of the reaction is slow, the potential reading drifts and the line becomes curved near the end point. However, the part of the curve farthest from the end point is a proper straight line. The points of this part are extrapolated back to the horizontal axis giving the volume required to reach the equivalence point.

Practically, the potential of the solution is measured as a function of the volume of the titrant. It is necessary, therefore, to convert the observed electrode potential to concentration. This can be done either by using a calibration curve or the log scale on a "Specific Ion Meter" or by using the known addition or subtraction technique. The concentration values as obtained by these methods must be corrected for the volume change, point by point, since the sample has been diluted by the titrant. Corrected values for the concentration C_c are obtained for the observed concentration C_0 by the relation:

$$C_c = C_0 \cdot V_s + V_t / V_s \qquad (36)$$

where V_s is the sample volume and V_t is the volume of the titrant added. This technique is tedious and it gained little attention until the special volume corrected graph paper (for obtaining linear plots citing this relation) became commercially available [70–72].

1. Gran's Paper

Although the relation between the concentration and the electrode potential can be plotted on semi-log paper to give a straight line as the potential varies linearly with the log of the ion concentration, the commercially available paper is based on the semi-antilog relation (Fig. 3.10).

Semi-antilog paper is constructed [73] with divisions on the antilog axis corresponding to E/S. A straight line is obtained on this paper by plotting the electrode potential vs ion concentration.

$$E/S = E_0/S + \log[X^{\pm}] \qquad (37)$$

$$\text{antilog } E/S = E_1 + [X^{\pm}] \qquad (38)$$

where E_1 is the antilog E_0/S. This semi-antilog paper gives zero concentration as a point on the calibration graph by extrapolating the straight line.

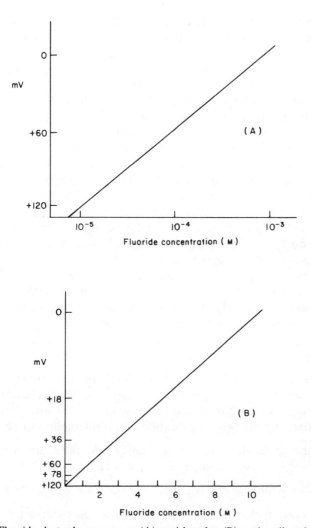

FIG. 3.10. Fluoride electrode response on (A) semi-log plot; (B) semi-antilog plot (according to *Orion Newsletter* 2, 49 (1970)). (Courtesy of Orion Research Inc.)

This allows a linear titration plot which can be extrapolated to the equivalence point. As the titrant solution is added, the concentration of the ion being sensed changes and the change in potential ΔE is an antilogarithmic function of concentration changes:

$$\text{antilog } \Delta E/S = [X_1]/[X_2] \qquad (39)$$

Orion's paper (90-00-90) is based on a 58 mV slope. By rescaling the vertical divisions by a factor of 2, the slope can be made 29 mV (for divalent ions). The paper is corrected for volume changes up to 10% of the original sample volume. This correction is accomplished by skewing the vertical axis upward by 10% from the left to the right-hand side of the paper. To obtain a straight line by this paper the total ionic strength of the titration solution must remain constant, the pH of the solution must remain within certain limits and a large excess of pH buffer solution must be employed, if this buffer acts as a complexing agent.

For the determination of an ion (cation or anion) using the ion-selective electrode and Gran's paper, 100 ml of the solvent containing a suitable reagent to give a constant ionic strength background is subjected to titration with the titrant which will be used in the true experiment. Total ionic strength adjustment buffers (TISAB) have been suggested [71, 74, 75] for this purpose. The titrant is added in aliquots of 0·5 ml until a total of 10 ml has been delivered. The millivolt reading obtained for each addition is recorded on Orion Gran's paper. A straight line passing through these points must be obtained and will intercept the horizontal axis at the origin, if the electrode slope is 58 mV and the level of the ion to be determined is very low in the background solution. The titration is repeated on unknown sample solutions [76–79]. A straight line passing through the last 4 or 5 points is extrapolated to intercept the horizontal axis.

Using Gran's paper, the following points should be taken into consideration [73].

(i) If the electrode slope is less than 58 mV for a monovalent electrode (e.g. 56 mV), the intercept will be slightly in error about one minor division on the horizontal axis. The amount of the error can be easily determined by running a calibration graph for the ion to be measured.

(ii) If the electrode slope is less than 58 mV, the extrapolation of the calibration graph will intercept the horizontal axis to the left of the origin.

(iii) If the slope is greater than 58 mV, the intercept will be to the right.

(iv) If a cation electrode is used, values must be increasingly positive up the axis.

(v) If an anion electrode is used, values must be increasingly negative up the axis.

(vi) If a monovalent electrode is used, each major division is equal to 5 mV and each minor division equal to 1 mV.

(vii) If a divalent electrode is used, each major division is 2·5 mV and each minor division is 0·5 mV.

(viii) If the ion being sensed by the electrode is a sample ion, the line will slant from the upper-left to the lower-center of the paper.

(ix) If the ion to be measured is the titrant ion, the curve will fall to the right of the equivalence point and will slant towards the upper right-hand corner of the paper.

Gran's plot offers the advantage of calculating the end point by titrimetric procedure in which the precipitate is relatively soluble, or in which the soluble part tends to supersaturate. Under these conditions, it is impossible to locate the end point by the conventional methods since it has been obscured by the onset of precipitate [80].

2. Gran's Ruler

The Gran ruler is a device which permits Gran plots of titration, standard addition and subtraction techniques on linear graph paper. This ruler has an inbuilt antilogarithmic function corresponding to definite changes in potential. Thus, any linear graph paper can be used and there is no need for the use of expensive semi-antilogarithmic paper. Volume changes are plotted on the abscissa and the cumulative potential changes are plotted directly antilogarithmically on the ordinate. A specific slope (S) and a volume correction are built into the paper or ruler.

An example of the use of this ruler is shown in Fig. 3.11. The ruler base is aligned with the horizontal axis and $O\Delta E$ is plotted on the left vertical axis. Successive potentials (cumulative ΔE values) are plotted at each volume increment. Volume correction is made by adjusting the baseline with the horizontal axis.

The ruler instructions [81] are as follows.

(i) On linear graph paper, draw a vertical (ΔE) and a horizontal volume axis.

(ii) Align the ruler baseline with the horizontal axis.

(iii) Fix the potential on the vertical axis in the following manner:
   (a) *Sensing sample* (*before end point of titration*). Set the ruler on the left vertical axis and plot the first potential opposite $O\Delta E$ on the ruler.
   (b) *Sensing excess of titrant* (*after end point of titration*). Set the ruler on the right vertical axis. Plot the last potential opposite $O\Delta E$ on the ruler.
   (c) *Standard addition.* Plot the last potential opposite $O\Delta E$ on the ruler at the right vertical axis.

(iv) Plot cumulative ΔE values *vs* volume as follows. Use the ΔE scale on the left-hand side of the ruler for monovalent ions and on the right-hand side for divalent ions. Volume correction is made by changing the ruler alignment relative to the horizontal axis. For each ΔE value plotted, move the ruler on one correction unit for each volume increment.

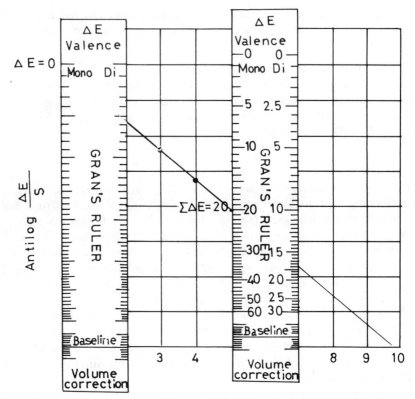

FIG. 3.11. Gran ruler (according to Westcott, *Anal. Chim. Acta* **86**, 269 (1976)). (Courtesy of Elsevier Scientific Publishing Co.)

With increasing volume (plotted from left to right) use method (iiia) and raise the ruler baseline. With decreasing volume (plotted from right to left) use either method (iiib) or (c) and lower the ruler baseline.

(v) Draw the best straight line through the plotted points and extrapolate to the horizontal axis. In methods (iiia) and (b) this indicates the titration end point. In the standard addition method, the number of volume units is between the zero addition point (U_0) and the point at which the plotted line crosses the horizontal axis (U_x) to the left. Multiply this value by the known standard concentration (C_s) and volume (V_s), then divide by the sample volume (V) to obtain the sample concentration (C).

$$C = (U_x - U_0)C_s V_s / V \qquad (40)$$

3. *Complexometric Titrations and Gran's Plot*
In complexometric titration, difficulties may arise in obtaining straight lines on Gran's plot paper, probably due to the formation of more than one

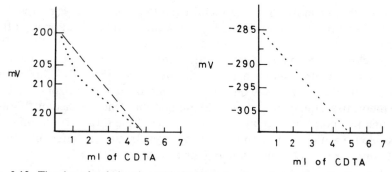

FIG. 3.12. Titration of cadmium ion with CDTA using the cadmium ion-selective electrode and Gran's plot in: (A) acetate buffer of pH 6; (B) citrate buffer of pH 6 (according to *Orion Newsletter* **3**, 17 (1971)). (Courtesy of Orion Research Inc.)

complex for some metals and the effect of pH. These factors have minor effects on the precipitation titration.

When the solution to be titrated contains a weak complexing agent, it must be present in large excess over the metal ion so that the extent of the weak complex of the untitrated metal ion does not vary as the titration proceeds. If the extent of weak complexation varies, curved lines will be obtained on the Gran's plot paper.

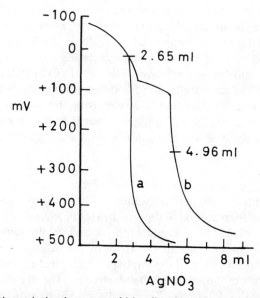

FIG. 3.13. Potentiometric titration curves of (a) sodium bromide, and (b) a mixture of sodium chloride and sodium bromide, titrated with silver nitrate using the bromide ion-selective electrode. (Courtesy of Orion Research Inc.)

Formation of more than one metal-ligand complex can be detected by using Gran's plot, but it is not readily apparent when the titration is plotted in the conventional manner. For example, titration of cadmium with CDTA (1,2-cyclohexylene diamine tetraacetic acid) in slightly acidic sodium acetate buffer of pH 6 shows a line with two slopes [82]. The line drawn through the last few points on the curve intercepts the horizontal axis at the theoretical equivalent, whereas extrapolation of the line passing through the first few points intercepts the half volume of the titrant, confirming the possible formation of 1 : 1 and 1 : 2 complexes. In citrate buffer solution, a 1 : 1 complex is only formed and one line is obtained (Fig. 3.12).

4. Precipitation Titrations and Gran's Plot

Titration of a mixture of chloride and bromide ions with silver nitrate using the bromide electrode can be represented by using Gran's plot paper [83]. Figure 3.13 shows the titration curve of a binary mixture of these ions. On Gran's paper draw linear plots for both the front side (the portion of the titration curve before the end point where the bromide ion is sensed) using increasingly negative values for the divisions which are increasingly high up the vertical axis of Gran's paper, and the back side curve (the portion after the end point where the silver ion is sensed) with increasingly positive potential away from the origin. Extrapolation of both lines leads to an intercept of the horizontal axis at the equivalence point (Fig. 3.14). By plotting both curves on the same volume-corrected Gran's plot paper, two separate intercepts are obtained. The first represents the bromide level and the second the total bromide and chloride (Fig. 3.15).

Since the end point is located by extrapolation from a region in which the level of bromide is high relative to the level of chloride, the tendency to form a mixed precipitate or solid solution is reduced, and the titre of bromide is less in error than that observed using the conventional titration curve (Fig. 3.13). A further advantage is that it is not necessary to take a lot of data near the end point; just a few points on the titration curve are sufficient.

5. Known Addition Technique and Gran's Plot

The known addition "spiking" technique can be employed using the Gran's plot paper. The increments of the ion being measured are added to the sample and the electrode potentials are plotted on the paper [73, 84]. A straight line is obtained and the intercept of the extrapolation of this line will yield the concentration of the ion being measured, plus a concentration equal to the limit of detection of the electrodes. The actual concentration is calculated by the difference. The results gained by this technique show better precision than those obtained from the conventional two points known addition procedure. The useful application of the known addition

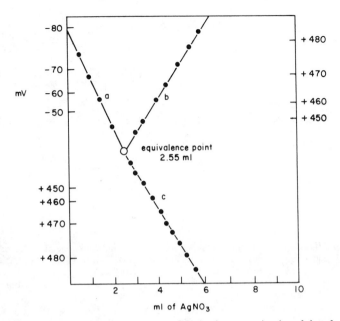

FIG. 3.14. Front and back titration curves on Gran's plot paper (replot of data from curves of Fig. 3.13) (according to *Orion Newsletter* **3**, 1 (1971)). (Courtesy of Orion Research Inc.)

technique using Gran's plot is very clear when the concentration of the sample to be measured is close to the electrode's limit of detection.

Gran's plots are applicable with automatic titration or in conjunction with computer systems for the analysis of the data obtained by the multiple addition technique [67, 72, 85–88]. Frazer *et al.* [89] have recently described

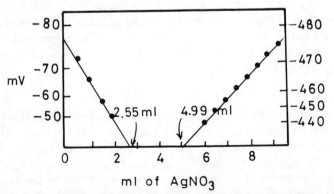

FIG. 3.15. Location of the equivalence points in the titration of chloride and bromide mixture with silver nitrate using the bromide electrode and Gran's plot (replot of data from curve (b) in Fig. 3.13). (Courtesy of Orion Research Inc.)

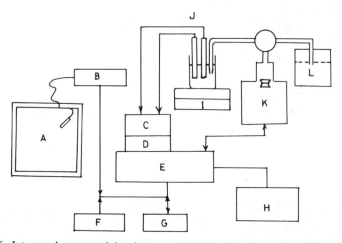

FIG. 3.16. Integrated automated titration system for use with the ion-selective electrodes and Gran's plot (according to Selig, Frazer and Kray, *Mikrochim. Acta* 1975, 675). (A) Oscillo-scope Tektronix type 611; (B) graf/pen; (C) data amplifier; (D) A/D converter; (E) computer PDP-8/I; (F) function box; (G) TTY; (H) plotter; (I) gold-plated stirrer; (J) electrode system; (K) Mettler digital burette; (L) titrant. (Courtesy of Springer Verlag.)

an integrated automated titration system based on a PDP-8/I minicom-puter. The computer controls a digital burette drive (Fig. 3.16), and can take a maximum of 512 data points. The data are stored in the computer memory and are displayed on the oscilloscope as a plot of e.m.f. *vs* volume of titrant, while the titration is in progress. After completion of the titration, the computer calculates and displays a Gran's plot. This system is applicable to both titrimetric and known addition procedures [84].

C. Null-point or differential potentiometric technique

In this technique, the concentration of the ion to be determined is compared with a solution of known composition using two identical indicator elec-trodes specific for the species to be determined. The composition of one of these solutions is adjusted to match the other [90–95]. The potential developed by this system is based on equation 41.

$$E = RT/n\mathrm{F} \ln C_1\gamma_1/C_2\gamma_2 + E_j \qquad (41)$$

where C_1 and γ_1 are the concentration and activity coefficient of the unknown solution; C_2 and γ_2 are the concentration and activity coefficient of the standard known solution; E_j is the liquid-junction potential. In the presence of a high concentration of an inert electrolyte in both half-cells, the activity coefficient of the ionic species in both cells will be approximately equal and the liquid-junction potential becomes negligible. Then:

$$E = RT/n\mathrm{F} \ln C_1/C_2 \qquad (42)$$

When the potential approaches zero, this means that both concentrations of the unknown and standard solutions are equal, $C_1 = C_2$.

The null-point potentiometry can be applied using various procedures.

(i) A standard reference solution is used as the null-point concentration solution and the concentration of the sample solution is varied either by titration or by addition of the same species to match the reference.

(ii) A known excess of the standard solution in a half cell is used in conjunction with the sample half-cell and the concentration of the standard is reduced by titration until the e.m.f. becomes zero.

(iii) A standard solution is added to a half-cell containing an inert electrolyte in conjunction with the sample half-cell until the e.m.f. becomes zero.

This technique offers the measurement of concentration, not from a single potential reading, but by titrating the sample with a standard to a potential corresponding to that obtained when both arms of the cell contain an identical solution. This can be carried out in a simple differential potentiometric cell like that shown in Fig. 3.17. As low as 6 p.p.b. of chloride can be measured by this cell [96].

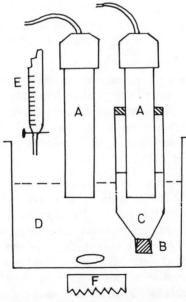

FIG. 3.17. Differential potentiometric cell (according to Florence, *J. Electroanal. Chem.* **31**, 77 (1971)). (A) Indicator and reference ion-selective electrodes; (B) ground glass sleeve; (C) reference solution; (D) sample solution; (E) burette; (F) magnetic stirrer. (Courtesy of Elsevier Sequoia S.A. Lausanne.)

Since the volume of one of the half-cells changes by addition of the titrant, a correction due to this dilution is applied:

$$C_2 = C_0 V / V_0 + V_a \qquad (43)$$

where C_2 is the concentration of the ion in the titration half-cell, C_0 is the titrant concentration, V_0 is the original volume of solution to be titrated, and V_a is the volume of the titrant added.

Graphical evaluation of the equivalence point is performed by plotting the e.m.f. *vs* titrant concentration on a semi-log paper and the equivalence point is obtained by a line interpolation to the null-point potential where e.m.f. is zero [94]. The titrant is added and the e.m.f. measurements are started when the cell potential changes to about 30 mV and taken at arbitrary increments of the titrant solution thereafter depending on the particular concentration range. A straight line with a theoretical Nernstian slope (59 mV per 10-fold change in concentration of a monovalent ion) should be obtained. Significant deviation from this value indicates improper cell behavior.

Coulometric generation of the titrant, *in situ*, at constant current has also been described [90, 91] to avoid a dilution effect. Also a series of standard solutions with concentration approximately equal to the unknown solution (a minimum of two solutions above and below this concentration) may be prepared and compared with the unknown [93].

Several important advantages are offered by the null-point potentiometry in addition to its intrinsic simplicity and high interference tolerance level. These are:

(i) Very small volumes of the unknown solutions can be used. The actual titration can be performed in any convenient volume of solution while the volume of test solution is unchanged since it serves only as a reference. The electrode itself can serve as sample container by using it in an inverted position and surrounding the membrane-end side with a Tygon tubing sleeve [94] (Fig. 3.18).

(ii) Very low concentrations of the unknown species can be measured since the volume ratio between the sample and titrant solutions acts as an amplification factor. This permits the titration of amounts too small to be titrated by the ordinary potentiometric procedures. For example, an unknown solution having a volume of 10 μl can be measured using a half-cell of a standard solution whose volume is 100 ml (i.e. the ratio is 1:10,000). The number of equivalents of the titrant added to the titration half-cell in this case must be 10^4 times greater than the amount present in the unknown half-cell in order to achieve the same concentration.

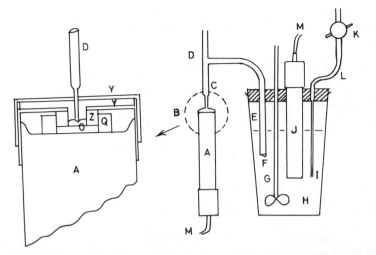

FIG. 3.18. Linear null-point potentiometric cell for the analysis of fluoride ion using the fluoride ion-selective electrode (according to Durst, *Anal. Chem.* **40**, 931 (1968)). (A) Modified fluoride electrode containing 0·1 M NaF, 0·1 M KCl and 4% agar gel; (B) sample microcell; (C) polyethylene capillary containing 0·1 M KNO_3–4% agar gel; (D) salt bridge with 0·1 M KNO_3; (E) Tygon tubing; (F) porous Vycor salt bridge junction; (G) stirrer; (H) variable fluoride solution; (I) polyethylene capillary tube; (J) fluoride electrode; (K) micro-burette; (L) titrant solution; (M) electrodes ends to pH-meter; (Y) polyethylene caps; (Z) polyethylene ring; (Q) Tygon sleeve; (O) electrode membrane; (U) sample solution (10 μl). (Courtesy of The American Chemical Society.)

(iii) The unknown sample solution is not modified or chemically reacted during the titration and can therefore be recovered unchanged for use in other measurements.

REFERENCES

1. S. Ebel, E. Glaser and A. Searing, *Z. Anal. Chem.* **291**, 108 (1978).
2. B. Fry and D. Taves, *J. Lab. Clin. Med.* **75**, 1020 (1970).
2a. M. Peters and D. Ladd, *Talanta* **18**, 655 (1971).
3. J. Ridden, R. Barefoot and J. Roy, *Anal. Chem.* **43**, 1109 (1971).
4. A. Gordievskii, A. Syrchenkov, N. Savvin, V. Shterman and G. Kozhukhova, *Zav. Lab.* **38**, 265 (1972).
5. M. Frant, *Plating* **58**, 686 (1971).
6. E. Pungor and K. Toth, *Analyst* **95**, 625 (1970).
7. *Orion Research Newsletter* **2**, 42 (1970).
8. Orion Research, Instruction Manual of Divalent Cation Electrode, 1st Edn (1966).
9. J. Andelman, *J. Water Pollution Control Federation* **40**, 1844 (1968).
10. D. Shearer and G. Morris, *Microchem. J.* **15**, 199 (1970).
11. J. Van Loon, *Analyst* **93**, 788 (1968).
12. R. C. Rittner and T. S. Ma, *Mikrochim. Acta* 1972, 404.

13. K. Yatsimirskii, *Kinetic Methods of Analysis*. Pergamon Press, Oxford (1966).
14. G. Guilbault, *Enzymatic Methods of Analysis*. Pergamon Press, New York (1970).
15. H. Mark Jr and G. Rechnitz, *Kinetics in Analytical Chemistry*. Wiley-Interscience, New York (1968).
16. G. Guilbault, *Anal. Chem.* **42**, 334 R (1970).
17. T. Neubecker and G. Rechnitz, *Anal. Lett.* **5**, 653 (1972).
18. D. Haisman and D. Knight, *Biochem. J.* **103**, 538 (1967).
19. P. Horowitz and F. DeToma, *J. Biol. Chem.* **245**, 984 (1970).
20. F. Duke, M. Weibel, D. Page, V. Bulgrin and J. Luthy, *J. Amer. Chem. Soc.* **91**, 3904 (1969).
21. H. Booker and J. Haslam, *Anal. Chem.* **46**, 1054 (1974).
22. L. Michaelis and M. Menten, *Biochem. Z.* **49**, 333 (1913).
23. G. Guilbault and J. Montalvo, *J. Amer. Chem. Soc.* **91**, 2164 (1969).
24. G. Guilbault, R. Smith and J. Montalvo, *Anal. Chem.* **41**, 600 (1969).
25. P. Alexander and G. Rechnitz, *Anal. Chem.* **46**, 250 (1974).
26. R. Llenado and G. Rechnitz, *Anal. Chem.* **45**, 2165 (1973).
27. R. Llenado and G. Rechnitz, *Anal. Chem.* **45**, 826 1973).
28. A. Altinato and B. Pekin, *Anal. Lett.* **6**, 667 (1973).
29. E. Baumann, *Anal. Chim. Acta* **42**, 127 (1968).
30. *Orion Research Newsletter* **1**, 21 (1969).
31. M. Frant, *Plating* **58**, 686 (1971).
32. J. Buffle, N. Parthasarathy and D. Monnier, *Chimia* **25**, 223 (1971).
33. T. Anfalt and D. Jagner, *Anal. Chim. Acta* **53**, 13 (1971).
34. A. Gyllenspetz, D. Kitchen and T. Ress, *Chemy Ind.* 1973, 640.
35. U. Hemerson, *Anal. Chem.* **43**, 1120 (1971).
36. J. Tusl, *J. Assoc. Offic. Agr. Chemists* **53**, 267 (1970).
37. H. Moeken, H. Eschrich and G. Willeborts, *Anal. Chim. Acta* **45**, 233 (1969).
38. L. Torma and B. Ginther, *J. Assoc. Offic. Agr. Chemists* **51**, 1181 (1968).
39. C. Martin and S. Brun, *Trav. Soc. Pharm. Montepellier* **29**, 161 (1969).
40. M. Smith and S. Manahan, *Anal. Chem.* **45**, 836 (1973).
41. G. DeBaenst, P. Van den Winkel and D. Massart, *J. Pharm. Belg.* **28**, 188 (1973).
42. J. Krueger, *Anal. Chem.* **46**, 1338 (1974).
43. J. Melton, W. Hoover and L. Ayers, *J. Assoc. Offic. Anal. Chemists* **57**, 508 (1974).
44. C. Rix, A. Bond and J. Smith, *Anal. Chem.* **48**, 1236 (1976).
45. D. Jagner and V. Pavlova, *Anal. Chim. Acta* **60**, 153 (1972).
46. *Orion Research Newsletter* **2**, 33 (1970).
47. R. Durst, *Mikrochim. Acta* 1969, 611.
48. R. Durst and J. Ross Jr, *Anal. Chem.* **40**, 1343 (1968).
49. M. Brand and G. Rechnitz, *Anal. Chem.* **42**, 1172 (1970); **21**, 1652 (1971).
49a. G. Horvai, L. Domokos and E. Pungor, *Z. Anal. Chem.* **292**, 132 (1978).
50. B. Karlberg, *Anal. Chem.* **43**, 1911 (1971).
51. G. Bronow, T. Ilus and G. Miksche, *Acta Chemica Scand.* **26**, 1117 (1972).
52. *Orion Research Newsletter* **2**, 26 (1970).
53. *Orion Research Newsletter* **1**, 35 (1969).
54. *Orion Research Newsletter* **2**, 1 (1970).
55. W. Selig and A. Salomon, *Mikrochim. Acta* 1974, 663.
56. S. S. M. Hassan, *Z. Anal. Chem.* **266**, 272 (1973).
57. J. Ross Jr and M. Frant, *Anal. Chem.* **41**, 1900 (1969).
58. E. Hakoila, *Acta Chem. fenn.* **B46**, 174 (1973).
59. E. Pungor, *Anal. Chem.* **39**, 28A (1967).
60. D. Katz and A. Mukherji, *Microchem. J.* **13**, 604 (1968).

61. B. Csakvari and K. Meszaros, *Hung. Sci. Instrum.* **11**, 9 (1968).
62. D. Küttel, O. Szabadka, B. Csakvary, K. Meszaros, J. Havas and E. Pungor, *Magy. Kem. Foly.* **75**, 181 (1969).
63. Y. Elmehrik, S. Marei and S. S. M. Hassan, *Libyan Journal of Science* **6**, 23 (1976).
64. W. Selig, *Mikrochim. Acta* 1971, 46.
65. F. Schultz, *Anal. Chem.* **43**, 502 (1971).
66. P. Carr, *Anal. Chem.* **44**, 452 (1972).
67. F. Schultz, *Anal. Chem.* **43**, 1523 (1971).
68. P. Carr, *Anal. Chem.* **43**, 425 (1971).
69. G. Gran, *Analyst* **77**, 661 (1952).
70. F. Rossotti and H. Rossotti, *J. Chem. Education* **42**, 375 (1965).
71. A. Liberti and M. Mascini, *Anal. Chem.* **41**, 676 (1969).
72. C. Coetzee and A. Basson, *Anal. Chim. Acta* **56**, 321 (1971).
73. *Orion Research Newsletter* **2**, 49 (1970).
74. M. Frant and J. Ross, *Anal. Chem.* **40**, 1169 (1968).
75. M. Peters and D. Ladd, *Talanta* **18**, 655 (1971).
76. W. Selig, *Mikrochim. Acta* 1973, 87.
77. T. Eriksson, *Anal. Chim. Acta* **58**, 437 (1972).
78. B. Fry and D. Taves, *J. Lab. Clin. Med.* **75**, 1020 (1970).
79. T. Takahari and M. Kosaka, *Japan Analyst* **25**, 192 (1976).
80. W. Selig, *Mikrochim. Acta* 1974, 515.
81. C. Westcott, *Anal. Chim. Acta* **86**, 269 (1976).
82. *Orion Research Newsletter* **3**, 17 (1971).
83. *Orion Research Newsletter* **3**, 1 (1971).
84. W. Selig, J. Frazer and A. Kray, *Mikrochim. Acta* 1975, 675.
85. I. Hansson and D. Jagner, *Anal. Chim. Acta* **65**, 363 (1973).
86. E. Ebel and R. Krommelbein, *Z. Anal. Chem.* **264**, 342 (1973).
87. E. Ebel and R. Krömmelbein, *Z. Anal. Chem.* **256**, 28 (1971).
88. T. Anfalt and D. Jagner, *Anal. Chim. Acta* **57**, 165 (1971).
89. J. Frazer, A. Kray, W. Selig and R. Lim, *Anal. Chem.* **47**, 869 (1975).
90. H. Malmstadt and H. Pardue, *Anal. Chem.* **32**, 1034 (1960).
91. R. Durst and and J. Taylor, *Anal. Chem.* **39**, 1375 (1967).
92. H. Malmstadt, T. Hadjiioannou and H. Pardue, *Anal. Chem.* **32**, 1039 (1960).
93. R. Durst, E. May and J. Taylor, *Anal. Chem.* **40**, 977 (1968).
94. R. Durst, *Anal. Chem.* **40**, 931 (1968).
95. J. Buffles, N. Parthasarathy and D. Monnier, *Chimia* **25**, 223 (1971).
96. T. Florence, *J. Electroanalyt. Chem.* **31**, 77 (1971).
97. L. Sucha and M. Suchanek, *Anal. Lett.* **3**, 613 (1970).
98. N. Radic, *Analyst* **101**, 657 (1976).
99. B. Jaselskis and M. Bandemer, *Anal. Chem.* **41**, 855 (1969).
100. E. Baumann, *Anal. Chem.* **42**, 111 (1970).
101. A. Homola and R. James, *Anal. Chem.* **48**, 776 (1976).
102. W. Bazzelle, *Anal. Chim. Acta* **54**, 29 (1971).
103. E. Hopirtean, C. Liteanu and R. Vlad, *Talanta* **22**, 912 (1975).
104. F. Chang and K. Cheng, *Anal. Chim. Acta* **76**, 177 (1975).
105. T. Anfalt and D. Jagner, *Anal. Chim. Acta* **56**, 477 (1971).
106. J. Ruzicka and E. Hansen, *Anal. Chim. Acta* **63**, 115 (1973).
107. M. Mascini and A. Liberti, *Anal. Chim. Acta* **64**, 63 (1973).
108. T. Hadjiioannou and D. Papastathopoulos, *Talanta* **17**, 399 (1970).
109. S. Tackett, *Anal. Chem.* **41**, 1703 (1969).
110. A. Hulanicki and M. Trojanowicz, *Chem. Anal.* **18**, 235 (1973).

111. W. Wood, *J. Res. U.S. Geol. Survey* **1**, 237 (1973).
112. K. Cheng, J. Hung and D. Prager, *Microchem. J.* **18**, 256 (1973).
113. E. Baumann and R. Wallace, *Anal. Chem.* **41**, 2072 (1969).
114. Orion Research, Analytical Methods Guide, 7th Edn (1975).
115. E. Hakoila, *Anal. Lett.* **3**, 273 (1970).
116. T. Tanaka, K. Hiiro and T. Kinoyama, *Osaka Kogyo Gijutsu Shikenshokiho* **21**, 93 (1970).
117. E. Moya and K. Cheng, *Anal. Chem.* **42**, 1669 (1970).
118. M. Taga, M. Mizuguchi, H. Yoshida and S. Hikime, *Bunseki kiki* **14**, 230 (1976).
119. G. Rechnitz and N. Kenny, *Anal. Lett.* **3**, 509 (1970).
120. G. Rechnitz, S. Zamochnick and S. Katz, *Anal. Chem.* **35**, 1322 (1963).
121. W. Selig, *Mikrochim. Acta* 1973, 349.
122. C. Botre, M. Mascini, B. Bencivenga and G. Pallotti, *Farmaco, Ed. Prat.* **28**, 218 (1973).
123. R. Levins, *Anal. Chem.* **43**, 1045 (1971).
124. J. Rousseau, *Analusis* **2**, 718 (1973/1974).
125. M. Frant, *Plating* **58**, 686 (1971).
126. S. S. M. Hassan, *Anal. Chem.* **49**, 45 (1977).
127. F. Conrad, *Talanta* **18**, 951 (1971).
128. P. Evans, G. Moody and J. Thomas, *Lab. Pract.* **20**, 644 (1971).
129. M. Frant and J. Ross, *Science* **154**, 1553 (1966).
130. J. Lingane, *Anal. Chem.* **39**, 881 (1967).
131. W. Selig, *Z. Anal. Chem.* **249**, 30 (1970).
132. W. Selig, *Mikrochim. Acta* 1970, 229.
133. W. Selig, *Mikrochim. Acta* 1970, 337.
134. S. S. M. Hassan, *Mikrochim. Acta* 1974, 889.
135. T. Light and R. Mannion, *Anal. Chem.* **41**, 107 (1969).
136. J. Orenberg and M. Morris, *Anal. Chem.* **39**, 1776 (1967).
137. H. Jacin, *Die Stärke* **25**, 271 (1973).
138. B. Hipp and G. Langdale, *Soil Sci. and Plant Anal.* **2**, 237 (1971).
139. D. Cantliffe, G. MacDonald and N. Peck, N.Y. Food and Life Sci. Bull., 1970, No. 3, Plant Science No. 2 (1970).
140. P. Pommez and S. Stachenko, Proceedings of 1970 Technical Session on Cane Sugar Refining Research, Oct. 12, p. 82 (1970).
141. R. LaCroix, D. Kenney and L. Walsh, *Commun. Soil Sci. and Plant Anal.* **1**, 1 (1970).
142. M. Fishman and O. Feist, U.S. Geol. Surv., Prof. Pap., No. 700C, 226 (1970).
143. M. Kawamura and T. Kashime, Kyoritsu Yakka Daigaku Daigaku Kenkyu Nemp., 11 (1970).
144. G. Rechnitz, M. Kresz and S. Zamochnick, *Anal. Chem.* **38**, 973 (1966).
145. W. Kriigsman, J. Mansveld and B. Griepink, *Z. Anal. Chem.* **249**, 368 (1970).
146. W. Kriigsman, J. Mansveld and B. Griepink, *Clin. Chim. Acta* **29**, 575 (1970).
147. N. Bottazini and V. Crespi, *La Chimica e L'Industria* **52**, 866 (1970).
148. W. Potman and E. Dahmen, *Mikrochim. Acta* 1973, 404.
149. W. Selig and G. Crossman, *Z. Anal. Chem.* **253**, 279 (1971).
150. E. Baumann, *Anal. Chem.* **40**, 1731 (1968).
151. R. Overman, *Anal. Chem.* **43**, 616 (1971).
152. S. S. M. Hassan, *Talanta* **23**, 738 (1976).
153. J. DiGregorio and M. Morris, *Anal. Chem.* **42**, 94 (1970).
154. W. Selig, *Microchem. J.* **15**, 452 (1970).
155. J. Brand and M. Smith, *Anal. Chem.* **43**, 1105 (1971).
156. R. Baczuk and R. DuBois, *Anal. Chem.* **40**, 685 (1968).

157. W. Selig and G. Grossman, Informal Report UCID-15623, Lawrence Radiation Lab., Livermore, California (1977).
158. W. Selig, *Mikrochim. Acta* 1970, 564.
159. W. Selig, *Mikrochim. Acta* 1976, 9.
160. R. Cummins, *Detergent Age*, March 1968, 22.
161. R. Geyer and H. Grank, *Z. Anal. Chem.* **179**, 99 (1961).
162. R. Geyer, K. Chojnacki, W. Erxleben and W. Syring, *Z. Anal. Chem.* **204**, 325 (1964).
163. A. Budd, *J. Electroanal. Chem.* **5**, 35 (1963).
164. R. Geyer, K. Chojnacki and C. Stief, *Z. Anal. Chem.* **200**, 326 (1964).
165. C. Harzdorf, *Z. Anal. Chem.* **262**, 167 (1972).
166. R. Jasinski and I. Trachtenberg, *Anal. Chem.* **44**, 2373 (1972).
167. R. Heistand and C. Blake, *Mikrochim. Acta* 1972, 212.
168. R. Reynolds, *Water Resources Res.* **7**, 1333 (1971).
169. J. Ross and M. Frant, *Anal. Chem.* **41**, 967 (1969).
170. J. Goertzen and J. Oster, *Soil Sci. Soc. Am. Proc.* **36**, 691 (1972).
171. J. Hicks, J. Fleenor and H. Smith, *Anal. Chim. Acta* **68**, 480 (1974).
172. W. Selig, *Mikrochim. Acta* 1970, 168.
173. W. Selig, *Mikrochim. Acta* 1975, 665.
174. L. Manakova and N. Bausova, *Zav. Lab.* **42**, 635 (1976).
175. E. Green and D. Schnitker, *Marine Chemistry* 1974, 111.
176. R. Naumann and C. Weber, *Z. Anal. Chem.* **253**, 111 (1971).
177. J. Slanina, E. Buysman, J. Agterdenbos and B. Griepink, *Mikrochim. Acta* 1971, 657.
178. L. Gruen and B. Harrap, *J. Soc. Leather Trade's Chem.* **55**, 131 (1971).
179. M. Smith and S. Manahan, *Anal. Chim. Acta* **48**, 315 (1969).
180. E. Siska and E. Pungor, *Z. Anal. Chem.* **257**, 12 (1971).
181. S. S. M. Hassan and M. M. Habib, *Microchem. J.* **26**, 181 (1981).
182. S. S. M. Hassan and M. M. Habib, *Z. Anal. Chem.* **307**, 205 (1981).
183. S. S. M. Hassan and M. M. Habib, *Z. Anal. Chem.* **307**, 413 (1981).
184. S. S. M. Hassan and M. M. Habib, *Analyst* (In press).
185. S. S. M. Hassan and M. M. Habib, *Anal. Chem.* **53**, 508 (1981).

SUBJECT INDEX